Writing Green

Advocacy and Investigative Reporting About the Environment in the Early 21st Century

by Debra A. Schwartz, Ph.D.

WRITING *Green*

Advocacy & Investigative Reporting About the Environment in the Early 21st Century

by Debra A. Schwartz, Ph.D.

Baltimore, Maryland
www.apprenticehouse.com

Project Manager: Natalie Joseph
Editors: Meghan Connolly and Shannon Morgan
Cover design by Thomas E. Korp

First printing
10 9 8 7 6 5 4 3 2 1

13-Digit ISBN: 978-1-934074-01-5
10-Digit ISBN: 1-934074-01-2

Baltimore, Maryland
www.apprenticehouse.com

For Karl Grossman and the White Russians

One of the reasons this [environment] beat is perpetually interesting is that it's the grandest train wreck of ideological, scientific, and financial interests imaginable.

Peter Dykstra
Society of Environmental Journalists
listserv, December 8, 1998

Half of environment journalism is having the story, half is having the credibility.

Len Reed
Science and Environment Editor
The Oregonian (Portland)

Table of Contents

Acknowledgements

Over White Russians made with Tia Maria, Professor Karl Grossman of the State University of New York's College at Old Westbury teased out the idea behind the research that led to this book. For that he is in the company of angels.

Also in the company of angels is Dr. Todd Steven Burroughs for the idea to turn the interviews from that study into the seminar-style discussion presented here. Next to them are Dr. Katherine McAdams, Dr. Judith Paterson, and Kent Silberman, from whom I am sure I have learned more than I realize. Everyday something they taught me comes into focus a little better, and I understand more than I knew before.

With their encouragement, I preened the wings of Tom Balyes, David Helvarg, Tom Meersman, Charles Pekow, Paul Rogers, and Dale Willman, veteran investigative reporters covering the environment who allowed me to interview them for my exploration into what might constitute investigative reporting about the environment as advocacy. Talking with them was like getting personally acquainted with characters in a Firesign Theatre routine or Monty Python's Flying Circus. As you read their stories, you may see that for yourself.

On the publishing side, my deep gratitude goes out to Loyola College in Maryland's Professor Andrew Ciofalo for requesting a book-length effort from me; to Dr. Kevin Atticks for his saintly guardianship; Natalie Joseph for managing this effort; and Meghan Connolly and Shannon Morgan for their editing. Were it not for their dedication to his manuscript, you would not be reading this now.

To all of you, thank you, thank you, thank you for sharing your richness with me. For giving of your time generously, for offering your experience fully, for facilitating learning, for the record.

— Debbie, 2006

Foreword

This book begins to fill a gap in the existing literature about investigative reporting. It can be used in several ways, including to explore how investigative reporting appears to be changing in non-technical ways, and to consider whether investigative reporting about the environment has had any influence on the process of investigative reporting in the early 21st century.

For the sake of brevity, only relevant excerpts from full interview transcripts appear in this book. All of the interviews took place between March 2003 and May 2003. Perspectives on radio, newspaper, television, magazine, and book-length journalism are represented from reporters contributing to each medium. Also included are samples of their work.

The stories they tell in their reports differ from the stories they tell about themselves. Here, the reporters interviewed talk about how they do what they do, what they get out of it, what covering the environment means to them, and what motivated them to do it as investigative reporters.

You'll hear from Tom Bayles, weekend editor, environment editor, and investigative projects reporter for the *Sarasota Herald-Tribune*; David Helvarg, independent multimedia journalist currently heading the Oceans Awareness Project, a Washington, D.C.-based non-profit organization he established; Tom Meersman of the (Twin Cities) *Star-Tribune*; Charles Pekow, freelance magazine investigative reporter; Paul Rogers, environment reporter for the *San Jose Mercury News*; and Dale Willman, president of Field Notes Productions in Saratoga Springs, NY who has reported for NPR and CNN.

Journalism derives its name from the French word *jour*, meaning "day." As defined by Columbia University mass communication sociologist James Carey, "journalism is a diary, a daybook, a record of the significant happenings, occurrences, events, and sayings during the life of a community." As such, journalism preserves not only the exotic, rare and sacred but also the common, ordinary, and repetitive events of life, Carey writes in the forward to *Journalism: A Guide to the Reference Literature* (Cates 1997, p. viii). Benjamin Bradlee, executive editor of *The Washington Post*, formally described journalism as a profession where reporters and editors are "pledged to approach every assignment with the fairness of open minds and without prior judgment. The search for opposing views must be routine. Comment from persons accused or challenged in stories must be included. The motives of those who press their views upon us must routinely be examined, and it must be recognized that these motives can be noble and ignoble, obvious and ulterior" (Webb, 1978, pp. 1-2).

The *New Webster's Dictionary of the English Language* (1981) defines journalism as an occupation involving reporting of the news and typified by "a type of writing ideally characterized by objectivity, but sometimes written to appeal to current public taste" (p. 520). In *The Random House Dictionary*, Stein (1978) specifies journalism as "the profession of gathering, writing, editing, or publishing news, as for a newspaper" (p. 490).

With publication of *Public Opinion* in 1922, news collection pioneer Walter Lippmann forwarded the concept of fair and balanced news reporting by suggesting that journalists follow the scientists' lead to produce objective research on which to build their stories. This idea was intended to raise the character of both the journalist and the stories provided, elevating them above articles that were based in rumor more than fact. As such, he attempted to locate a method for journalists to use for approaching the truth, and reporting it. Lippmann (1931) described journalism as practiced by "untrained accidental witnesses" (p. 170). His effort in part sought to improve the quality of journalism in the United States and also prevent a resurgence of yellow journalism, a type of gaudy reporting dominant in the late 1800s revolving first around sin, sex, and violence, and second around accuracy (p. 170). Paneth (1983) further defines yellow journalism:

> To the familiar exaggerations of sensationalism, yellow journalism added the elements of misrepresentation and falsehood – scareheads printed in huge black or red types, faked pictures and stories, reckless editorials, the superficiality of the Sunday supplement. (p. 527)

Lippmann decided that academic training in journalism would improve the quality of reporting and editing, news selection, and of the stories themselves. The scientific process he advocated involved basing reports on direct observation rather than hearsay, for instance. In *Liberty and the News,* he wrote, "There is but one kind of unity possible in a world as diverse as ours. It is unity of method, rather than aim; the unity of disciplined experiment." The field of journalism, he wrote, ought to make its cornerstone the study of evidence and verification (Kovach and Rosenstiel, 2001, p. 73).

Nearly 100 years later, it seems more is needed. But what? The reporters interviewed here share their techniques, approaches, and reasons why they put them in practice. Their answers all stem from questions about advocacy – or to put it another way, fairness, accuracy, and balance.

Introduction

In the last 38 years, environment reporting in the United States has developed from an obscure beat to an established part of American and world mainstream journalism. In the last six years, the environment has become a societal value. It wasn't before. Today, it is generally accepted that everyone is entitled to clean air, clean land, and clean water, just like it is generally accepted that crime is bad, education is good, and businesses should make money.

Until the late 1960s, coverage of the environment mostly consisted of conservation writing – that which focused on habitat and open land preservation. The mass media and the public accepted pollution as part of industrial society during that time. But in 1970, the Environmental Protection Agency was established to research, monitor, and enforce environmental laws and issues. Some newspapers had reporters assigned to cover the environment a couple of days a week at that point. But with the advent of the EPA, editors began to think about the shadowy background bureaucratic oversight and law enforcement enjoys in this and other countries.

They envisioned juicy stories filed with intrigue, wrongdoing, and underdogs: the stuff that sells newspapers. So some assigned a full-time watchdog.

Now, living with severe pollution in the name of progress is not the foregone conclusion it was during the Industrial Age and throughout the 1900s. The question that continues to rage is: How clean is clean?

In 1994, about a year after the Society of Environmental Journalists formed, I joined. Throughout the 1990s, many members vented about colleagues who made fun of the environment beat. On an early listserv, they told of political, crime, business, and education reporters who would mock them, sometimes with a tease like, "Hey, Randy, saving the environment today?"

Venting online that year, Randy Edwards, environment reporter for the *Columbus Dispatch* in 1998, wrote:

Apparently it's acceptable to advocate on behalf of good government or against corruption, but if the stories with impact lead to tougher environmental laws or protection of a valuable natural resource, suddenly the reporter is a 'flannel-clad tree-hugger.'"

That year, Mike Mansur, environment reporter for the *Kansas City Star*, said he carried two business cards because he feared industry representatives would think he sided with environmental groups if he identified himself as an environment reporter. Also in 1998, a book titled, *Green Ink* by environment reporter and journalism professor Michael Frome hit the shelves. SEJ members who read it empathized with the stories he told. Three cases in particular took my attention as I read the book many years later, remembering parts of the stories as they happened, because the people they happened to wrote snippets on the listserv about what they were encountering.

Those stories are the personal experiences told by Phil Shabecoff of *The New York Times*; Kathy Durbin of *The Oregonian*, and Richard Manning of *The Missoulian.* All discussed what they did to write investigative reports about the environment. All of them had followed the strict rules outlined in "The Paul Williams Way," the approach endorsed by Investigative Reporters and Editors (IRE). It is the traditional model for investigative reporting, which begins with a premise the reporter sets out to prove. It involves piecing together a trail indicating evidence leading back to whoever it was who made the decision to make the bad thing happen. That method is known as "The Paul Williams Way" because it is the way Williams, founder of IRE, described in writing his process of managing investigative reports as they progressed. Shabecoff, Durbin, and Manning followed that method in the 1990s, but were accused of advocacy anyway.

With that sort of thing happening, SEJ members began innovating, coming up with new ways to protect their credibility that the old ways didn't seem to cover. Some of the roots they laid down from their experience, and the habits they adopted for their own personal style, are offered here for your gain.

About Journalists and Journalism

The reporters interviewed offered some generalizations as context for describing what they do. For instance, Tom Meersman, of the (Twin Cities) *Star Tribune* said it is a reporter's task to not only quantify things, but also qualify them.

"Qualify means to put them in context," he said. "Why do people care about the land? It's basically a value story. And also the issue of respect. Can motorized use of public lands co-exist with traditional recreational uses of the land, such as hiking and fishing and camping and cross-country skiing? In certain areas, nobody wants those areas any more for those types of purposes if they've been invaded by motorized vehicles, because it's noisy, it's dangerous, it's unpleasant. The story is much more about values and how do these machines and their increasing popularity fit into the quality of life. The quality of recreational opportunities that the state has traditionally had and now, in some areas, are under siege due to lack of control."

A journalist must not only provide context, but also educate, said Tom Bayles of the *Sarasota Herald-Tribune*. "I am motivated by opportunities to learn new things about the world around me," he said. Broadening that, Bayles added, "The motivation of any journalist is to make something better for a group of people who need help righting a wrong in the government, which might be people in Montana paying [to put] sand out here on Siesta Key. It's not really right.

"The best journalists are the ones who constantly learn. They love the profession because they are learning all the time."

With regard to journalistic privilege and journalistic integrity, Bayles stood firm. "Journalists don't help out the authorities," he said.

"But there are moral imperatives that come before journalistic goals. Your duty as a citizen in most cases comes before journalism. For instance, when I was doing the fire series (an immersion investigative report, where Bayles became a firefighter and battled raging forest fires to produce first-hand articles about the experience), there was an arsonist running around the woods lighting most of the big fires we covered that year. And I started having a pen pal through the online journal I did for the [*Sarasota Herald-Tribune*] web site while I was in the woods. He sounded awfully suspicious from things he was telling me – things that maybe only the person lighting the fires or the firefighters could know. So I thought he was former firefighter, or that he might be the firebug. So now I had a name that I should probably turn over to the authorities, but I'm a journalist and we don't help the authorities. What do I do, right?"

Bayles and his editors agreed it was his duty to stop the fires from happening. The police had not come to the paper and asked for information, and Bayles had gotten the information as a reporter, not as a private citizen. Bayles and his editors decided he would write an online diary that included a passage noting he had received a lot of e-mail on the topic. In his posting, Bayles quoted some of his suspect's correspondence, among others.

In another case, Bayles the reporter – not Bayles the citizen – saw a man starting a controlled burn that got out of hand and burned down a home. The authorities were investigating, but didn't know Bayles was on the scene working on his series. Again, he opted to write about it in a column in his online diary.

Even though reporters and editors know that kind of integrity is alive and well in journalism, overall journalists are misperceived, said Paul Rogers of the *San Jose Mercury News*. "A lot of reporters are iconoclasts," he said. "As the world changes around them, sometimes they don't change so well. But in the end, they are proud of the career choice they took.

"Also, journalists aren't cynics. They're actually idealists. They are people who believe in the power of the printed word, who believe in the power of ideas, who believe that if you present the public with all the facts, the public will make the right decision. And there's something wonderfully idealistic about that.

"It's the idea the whole county is premised on. People forget that. I have faith that common people without perhaps a level of wealth or education can better decide in a more evenhanded or free and just and compassionate way their own affairs. Journalists, deep down at their core people will want to do the right thing. And that's a very idealistic world view.

"Journalists also believe in the validity of institutions. If something gets screwed up, then they'll fix it. They believe that the free exchange of information makes people freer, and makes them richer, too."

An environment reporter's primary responsibility is to help mankind and the environment, Bayles said. NPR and CNN Radio reporter Dale Willman, president of Field Notes Productions in Saratoga Springs, NY, added that a reporter's obligation is to the audience. He said, "The environment beat isn't alone in this, but many of the stories we do can have a huge effect on people's health and their lives. So, if we're doing a story about PCB contamination in the Hudson [River], or lead in water supplies, or toxic releases from power plants, all those issues have potential ramifications in people's lives. And that's a hug responsibility, to get it right, to be as accurate as possible and say it in a tone that doesn't make it too alarmist and yet doesn't play down potential threats. So it has to strike a perfect balance between those, so you don't lose listeners [audience]."

Generally speaking, there are two types of environment reporters, Rogers said. "The one who likes to write about why things happen, and the type who likes to write about how things happen. 'How' people are interested in the mysteries of nature and science and how things work. They love to go out on boats with scientists and look at test tubes and ask what the levels are of pollutants.

"If they go into a forest and write a story, they'll say what the natural processes are there and how well they are working. They do a lot of explanatory journalism, and tend to be a little more scientific-minded," he said, adding Marla Cone of the *Los Angeles Times* is his favorite reporter of this sort.

"Why" reporters like to explain why things are the way they are, Rogers said. "Instead of going out on the boat with scientists and looking at test tubes, they would write about why those invasive species are out there. And why that pollution is there. Well, it's there, maybe, because this oil refinery gave a campaign contribution to the governor. They're less interested in the science.

"While the beat is a mix of politics and science, the 'how' writers are two-thirds science and one-third politics, and the 'why' writers are two-thirds politics and one-third science. 'Why' writers love the skullduggery of whose giving what money to whom."

Before joining the ranks of reporters covering the environment, multimedia reporter, David Helvarg, who established the Oceans Awareness Project in Washington, D.C., said he "noticed that most political and investigative reporters tended to view the environment beat as either uninteresting or low-prestige, while many environment reporters came to the beat from a science or nature writing background." That remains true today.

Bayles and Rogers further diced and sliced types of reporters into categories. Rogers said, "There are two types of investigative reporters: deep divers, and grazers. Deep divers like to find one interesting issue or piece of scholarship and dig in like scientists. Grazers prefer to sample everything. I am a grazer. Put me in a cubicle for six months to study a database and it breaks my stride.

"I'm glad I did that kind of investigative reporting," Rogers said, "but when I was in the middle of [it] I wanted to quit. Some reporters descend into an absolute funk while they're doing one project. They get sick of the topic. After about three weeks you normally don't want to ever see the topic again that you're writing about. It's so exhausting."

Bayles is of the other sort: a deep diver. He counts that type as not only immersion journalism but also explanatory journalism, which, he notes significantly, "doesn't necessarily have a balance. It's an explanation of what's going on. For instance, in the 'Inside the Fire' series, I was *embedded* with firefighters. The story is showing an explanation of what happens to the men and machines that fight the forest fires during a big fire year, back in the woods where no one ever sees them [for instance, metal machinery melts sometimes because of the high temperature]. And here's what it's like when people flee from a neighborhood that a fire pounced on, and burned down their homes. Here's the chaos. Here's an explanation of what it's like to be a firefighter, what it's like to stop a huge wildfire, what it's like to be a homeowner threatened by a wildfire. There's no greater purpose to that except to illuminate people's knowledge about something that hasn't been written about in that kind of detail.

"That's an investigation because it takes a lot of original reporting, of looking into things that aren't necessarily in a database anywhere that has been created by some government entity. It's the softest form of investigation for sure. But it's definitely going out to learn something – to investigate what it's like."

Helvarg, author of two works of book-length journalism about the environment, also counts as a deep diver. From his investigations, he gleans right-at-the-bone details he uses as a fine artist uses shadow and light to display a view. This is nowhere more evident in his work than in *The War Against the Greens*, an investigative effort about the wise-use movement that employs fiction techniques to herald a wake-up call.

Chapter 1: Tom Bayles

In elementary school, Tom Bayles received report cards saying he needed a lot of help with writing, prompting him to drop his first novel behind the couch at age 11. Today he is weekend editor, environment editor, and an investigative reporter for Sarasota's *Herald-Tribune* newspaper.

At the age of seven, the movie *All the President's Men* exposed Bayles to investigative reporting for the first time. His interest in the environment and wildlife developed when he became a Boy Scout. It grew as he swallowed almost whole the children's book *My Side of the Mountain* and its sequels, which he read over and over again.

At the University of South Florida, Bayles fell in love with having a byline while writing for the student newspaper as an undergraduate studying education. Mesmerized and fascinated by the atmospheric reactions between the earth and sky, Bayles started a master's degree in geomorphology at the school, intending to become a weatherman. But he craved a byline and shifted to gaining a master's degree in journalism instead.

His investigative reporting skills deepened during

two years with *The Associated Press.* At the *Herald Tribune*, they flourished when Rosemary Armao, former director of Investigative Reporters and Editors and managing editor for the paper until late 2002, drew him deeper into investigative reporting.

To get closer to a story about wildfires in Florida, Bayles became a firefighter. The result was an immersion report netting him a Pulitzer Prize nomination. When the *Herald-Tribune* installed a cable studio in a corner of its newsroom, Bayles made the jump to broadcasting but kept his newspaper positions. To prepare for on-camera work, he spent $800 on clothes and $100 on make-up (so he would "look good").

Bayles is a two-time Pulitzer Prize nominee for investigative reporting about the environment. He also holds assorted awards for feature writing, environmental reporting, and investigative reporting, including the Florida Society of Newspaper Editors Best of the Show gold medal award.

Inside the Fire: The story of the 2001 wildfire season in Southwest Florida
Part One

Published March 02, 2002 in *The Saratoga Herald*
By Tom Bayles

The arsonists drive into the North Port woods in the early afternoon, touching a lighter to the underbrush along Pan American Boulevard.

Most likely teen-agers on a lark, the kids might have a tough time getting things cooking this March day, if not for the drought gripping Florida.

The place they pick is a dried-up pond in a small stand of woods bordering a subdivision.

Water usually stands three feet deep at the spot, but now any moisture has sunk seven or eight inches into the parched soil.

The underbrush, like scrub all over Southwest Florida, is brittle. It leaps at the fire's touch.

Inch-high baby flames climb up palmetto bushes, ranging over dead lower limbs, mosses and vines, racing up 40-foot trees. A forest fire is off and raging, headed straight for a 110-bed nursing home.

If the 20 mph winds shift -- and winds shift often during a wildfire, steering it, feeding it -- and the fire moves left, a community center would be torched. If the fire moves right, at least six houses could be leveled.

The Pan American blaze is a typical Florida firestorm: Burning right next to homes, businesses, a community that hasn't been there very long.

Dark smoke billows as the fire eats through the woods. The wildfire crackles with a sound like a tornado, maybe more ominous.

The swirling heat rises from the wildfire, accompanied by the woosh of hot wind sucking up twigs and dirt.

The first 911 calls bring out city firefighters, men and women who know how to fight house fires, but not this kind of fire.

They park their engines on a strip of grass, all that stands between the blazing woods and the nursing home. Red-hot embers fall on the roof as the hallways begin to fill with acrid smoke and an evacuation begins.

An old woman stands against her bedroom window, her hands and face pressed to the glass, eyes large, frozen in place as she watches the wall of flames come at her.

Realizing they are outmatched, the firefighters call on the state Division of Forestry. Its tractor-mounted rangers are experts at fighting fires in dense Florida wildland where fire engines can't go.

All the while, the rangers have been listening to the radio, itching for an official request for help. Todd Ernst, Jeff Pilotto and Brian Olsen know that before they can fight the fire, they have to get the call. And then they have to get there.

They move their lumbering tractor-transports through back roads. Few motorists notice the emergency lights on the 18-wheelers, but that doesn't matter much. Even stretched to the limit by eager rangers, the heavy machines can't move much faster than cars doing the speed limit.

As he pushes his rig toward the fire, Ernst reaches for his cell phone to call Jenny. A hulking 6-foot-4, 265-pounder, Ernst always calls Jenny. Every fire. Just in case. They've been married five months.

Ernst unloads his tractor away from the other rangers, between the fire and the nursing home, intending to cut off the blaze. He'll do that by tilling a line in the woods with the plow Florida's forest rangers pull behind their tractors. His tractor is older, the cab open to the air, so limbs scratch him and smoke chokes him as he works.

Pilotto and Olsen drive newer tractors, ones with enclosed, air-conditioned cabs, though it's hardly a comfortable ride. They have eased their rigs into the woods, at the rear of the fire, on the scorched ground where the flames started. It's a safe spot; a fire can't burn what it has already consumed.

Five feet into fighting the fire, Ernst snarls his older-model tractor in a cable television wire running from a telephone pole to the nursing home.

It's going to be that kind of day.

Wind shoves the fire left, where Olsen and Ernst are cutting fire lines. Ernst moves in from the front, Olsen from the back. The plan: Meet up and create a continuous firebreak that protects the community center like a dry moat.

Not even 30 minutes old, the wildfire is now 50 acres big. A column of smoke rises above it, not only touching the clouds but becoming one.

And then, as if the ground has opened beneath them, Ernst and Olsen and their machines topple from view.

None of the firefighters saw the dried-out creek, with its sharp banks and overgrowth. The two tractors are stuck at 90-degree angles, straight down and about a football field apart.

The rangers try frantically to rock the tractors backward, up and out, up and out, up and out.

The incline is too steep.

"I'm trapped," Ernst yells into his radio. "I can't get out!"

As soon as Pilotto sees Olsen go over the edge he gets to work, hooking heavy chains between his 8-ton tractor and Olsen's. He urges his tractor full forward. Olsen drives just as hard in full reverse. But his tractor blade hangs up on a tree, the metal treads spinning deep into the sandy soil.

The fire is sweeping in. The heat is blistering, the roar of the flames hurts their heads.

At last, the tree gives, bending slowly sideways, and Olsen's tractor pops free.

Ernst is still in dire trouble. A flaming wall is coming at him at 200 yards a minute. He is cut off.

He had been having fun, the kind of fun grown men have taking big tractor boy-toys out to knock over trees, dig in the dirt, save homes.

But the air is scorching hot now, and Ernst is afraid. He is breathing in smoke. The tractor isn't budging.

He hears himself scream.

"I'm bailing out; I'm bailing out," he yells over the radio.

The radio. Ernst focuses on it, remembering his training. In a burnover, save that expensive piece of equipment, they teach you.

He tries to tug it from its mount. Tree tops 60 feet over his head are exploding. The damn thing won't shake loose.

The heat penetrating his helmet is approaching 1,200 degrees. The back of his neck starts to burn. That's it. Forget the radio and run.

He runs as fast as he ever has, down the sandy fire line he has just plowed, tripping over roots, toward the grassy field by the nursing home.

The wildfire overtakes the John Deere. Hydraulic and gas lines burst. The paint bubbles. The black seat Ernst has just fled melts like a child's crayon left on a sunny sidewalk. The tires incinerate into black ash.

But Ernst tumbles out of the woods, alive, unburned and furious. The fire has beaten him, forced him to run. He breaks into tears, throws his helmet to the ground, coughs out smoke.

It takes several hours and six drops of fire-retardant

slurry from federal firefighting planes to halt the blaze. The nursing home is saved; so are the community center and the homes.

When Pilotto and Olsen check on Ernst he is sucking oxygen inside an ambulance. His tractor is lost, his pride damaged.

But Jenny will see him at home tonight.

Firefighter Todd Ernst

Todd Ernst, 32, is a former restaurant manager who took a $25,000 cut in pay two years ago to go to work for forestry. He plans to ply the waters of Charlotte Harbor as a fishing guide in the wildfire off-season, snagging snook and redfish. Even at his young age, Ernst is a grandfather. His wife, Jenny, has two children from her first marriage, including a 21-year-old son with a 1-year-old son. Ernst thinks being a grandfather is cool.

Firefighter Jeff Pilotto

Jeff Pilotto, 33, spent four years in the Air Force, then nearly a decade with the Army National Guard when living in Fort Myers. In only his third year in forestry, he is one of the men who supervisors turn to first when a big fire breaks out. While springtime wildfires rule his work life, his private time is wrapped up with Courtney Shindle. After three years of dating he plans to ask her to marry him if he can find time during the fire season. He considered dressing up in the Smokey Bear costume to pop the question.

Update

Todd Ernst went fishing in Charlotte Harbor often over the winter, but has yet to get his captain's license so he can become a fishing guide. He finished near the top in several tournaments over the winter.

Jeff Pilotto made a surprise proposal to Courtney Shindle at a party she thought was to celebrate her birthday. They plan to marry this month.

Ed Vuolo continues to work with ranger Ray Sayer in Punta Gorda.

Frank House continues to smoke too many cigarettes and drink too much coffee.

Rick Christman is still in charge of the Punta Gorda Work Center.

Chuck Johnston returned to Sarasota County this year as its first controlled burn specialist. The position was created in hopes that never again will a contractor mess up and burn down a neighborhood as Mitchell Welch did.

Steve Wilson, an amateur paleontologist, went to the Maryland coast in the fall to search for shark fossils.

Ed Flowers, Arthur Coulter and Andrea Lee traveled to the Smoky Mountains of Tennessee and Virginia to battle wildfires in November.

Burn contractor Mitchell Welch settled lawsuits in December filed by homeowners and by the state for firefighting efforts resulting from the Carlton blaze. Terms of the settlement are confidential.

Editor's note: Herald-Tribune *reporter Tom Bayles trained to become a wildland firefighter. He spent the 2001 fire season on the front lines, fighting fires and documenting what he saw and experienced.*

Firefighters know that the big fires draw all the attention: The sightseers, the television reporters and their satellite trucks. But firefighters also know that the small fires can be the most hazardous. Ranger Ed Vuolo found out in May 2001 just how dangerous they can be.

Inside the Fire: The story of the 2001 wildfire season in Southwest Florida
About the series

Published March 02, 2002 in *The Saratoga Herald*
By Tom Bayles

Wildfire always has been a necessary part of the Florida landscape. For centuries, periodic fires have rejuvenated the state's inland forests and grasslands, clearing old brush to make way for new growth.

So long as Florida's population was clustered tight along the coast, the fires bothered no one. But in the course of the last decade, the state's population has grown by more than 3 million. As growth seeped inland, wildfires began to threaten, then consume homes and businesses.

During the last three years, Florida wildfires have charred more than 1 million acres, burned more than 560 homes and caused nearly $1 billion in damage. Wildland firefighters from nearly every state have come here to battle flames.

Just as there is a hurricane season in Florida, there is a wildfire season. It begins in late winter and continues through the spring, the dry months preceding the summer rainy season.

Last year's fire season began under particularly dangerous conditions, as the state entered another year of extended drought.

With that in mind, Herald-Tribune reporter Tom Bayles attended wildland firefighter training school to earn federal credentials needed to serve as a volunteer firefighter with the Florida Division of Forestry. He spent six months helping to fight some of the state's biggest wildfires while also watching and recording the events around him.

In his reporting, Bayles learned what it is like to be inside the fire. His story begins in North Port, on an early March afternoon in 2001.

Tom Bayles on Environment Reporting

Q: What drew you to environment reporting?

A: "I love to be out on the water. I love the beach. I love the forest. I wouldn't say I am an environmentalist, but I definitely love being outdoors and learning the cool things about animals and the reactions between the Earth and the sky."

Because of his love of the outdoors, Bayles is curious about geomorphology, the study of things that live in and things that occur one mile above the Earth and one mile below it.

"It's so cool. You've got volcanoes, plate tectonics, soil, water, all that kind of stuff – everything on the surface of the planet. Plus you can do lower stratosphere and weather patterns. So I knew that I loved to learn about that stuff – this is great stuff and Wow! There's a beat dedicated to all this.

"I had a friend, a reporter, and I was just so jealous because she got to write about the coolest stuff. I didn't know she had to go to water board meetings and sit through them, but most of what she got to write was about really significant things. I thought that was cool, and I made it a career goal to work myself into an environment beat if one should ever open up while I'm around."

Q: What personal lessons have you learned from environment reporting?

A: "[Environment reporting] has reinforced my desire to continue learning throughout my life because reporting on the environment is like a big science project. Quite often you have to learn new things about nature that keep feeding the brain. It reinforced my love of learning. It was nice to know that being on the environment beat is like being in college again, because it's rewarding in terms of learning, and getting paid for it."

Q: What makes the environment beat meaningful for you?

A: "It's really important that someone outside the government and special interests is watching what's going on with our environment. Just like if we screw up our schools, we screw up a whole generation of our children. If we screw up the environment, we screw up our planet. If we're not paying attention, we could all die. The planet could become uninhabitable. I'm going a bit far, but it's important that we in the media keep watch. That's right on, and key to how I feel about my job."

"I love the work. I even do it on my own time. Journalism is the greatest profession in the world. I get paid for something I would do for free. And that's really cool. And I get paid well. I have good benefits, and bosses treat me right. I have a lot of autonomy. Investigative reporting is the best because you have the greatest potential to change the world for a little bit better. It truly is the Fourth Estate, and I'm part of it.

"There's no way that I could have as much positive influence on my community except through journalism. I would never be a politician, I would never be an actor or any kind of activist or some great author that would change the world for the better. This is the only way I have to leave the world just a little bit better than I left it, in the smallest way, even if it's only Sarasota County rather than the whole United States or the world. If there are enough [people] out there who are inspired by this and we put some better rules in place to make the environment, to make it a little bit easier for humans and animals to co-exist, from both viewpoints, then there you go! That's as great a legacy as anybody can have – where I come from anyway.

"If you cover the city council and report what happened, that's great – people need to know that. They wouldn't go to the city council meeting, most of them. So that's an important service. But it's not going to change anything unless some major blow-up happens and someone gets canned.

"But with investigative reporting, you can pour your heart and soul into it and definitely have some positive outcomes: if not changing laws, then you're alerting society to potential problems that could be fixed. There you have it!"

Q: How do you view investigative reporting?

A: "Do you know what an investigative reporter needs to be? A master reporter. An investigative reporter has to be able to do any kind of reporting at a high level. When you were a city government reporter, you should have been a really good one. When covering cops, you should have just kicked ass. You know how to approach the school board. You know how to approach a cop. You know how to approach business. You know?

"And the little things that you as a reporter do to get into the mayor's office all the time. Make them your best friends. When you're the investigative reporter on a big project, all of that comes into play. And if you never were a schools reporter and never covered the environment and all of a sudden you're on a project about that, you're at a huge disadvantage. So a master reporter and a master generalist are your main keys to being a successful investigator.

"No special formal education is required. But I do think you're going to have a hard time being an investigative reporter unless you have some type of advanced degree. The higher level analytical thinking skills that you flex when you are in college come very much into play. I constantly compare what I do in my investigation with a very large term paper, trying to impress the hell out of your professor and get an A. It's the same thing – with perhaps a little more at stake. And you have more resources to help you get there, like a librarian and a budget.

"Investigative reporting is one of the lynchpins of the profession. It's what makes journalism the fourth estate, what makes it the watchdog for truth, justice, and the American way. The qualities of an investigative journalist are tenacity, organization, respect, humility, an incredible ability to clarify a complex situation, and absolute perfect fact-checking.

"There's nothing like seeing your byline on an investigative report. The feeling it gives. I thought it would be really cool to be an investigative journalist. That's the crème de la crème. It's fun, you're respected in the newsroom, and you do good things and you change the world. That's what I thought it would be, and it is that way. It's very cool.

"I love it. It's really cool to [hear that] you don't have to file a story for the foreseeable future – just go. Not only is that cool to be relieved from the daily grind, but more importantly it is a very heart-warming testament to your boss's belief in your ability to get the job done and be a self-starter. Because not everyone is a self-starter. Some people could not do investigative journalism. They'd be lost. [In some cases] they can't do them because maybe you don't see your byline for a year while you're doing it. That's tough."

As President George W. Bush deployed members of the armed services to Iraq, Afghanistan and elsewhere in the Middle East, Bayles was pulled off investigative project work to write stories with military connections throughout Florida as well as locally.

"It was tedium. I was punching out daily stories about the environment just to see my byline and stuff – they wanted that [environment stories] too – and I was getting very frustrated because here I was being a daily journalist again. I wasn't – I'm by no means saying that is beneath me – not at all! But it was just like (groan) I want to get back to a project! (Groan) I want to get back to a project! I'm happy to be even on a small project again. I like to be on projects, and now that I have a taste of blood I want more. I enjoy the hell out of it."

Q: What criteria do you use for moving the public with your investigative reports?

A: "There are three things an investigative report must have to be effective. The first is voice, the voice of authority. I'm still learning how to do that. It is a sense of confidence with the prose. It is not wavering on driving home your point that something needs to be fixed. People should not be in avoidable accidents. Period. Not, 'People shouldn't be dying in avoidable accidents, but you know the dredging companies really need to make some money and oh, they're poor dumb people anyway and maybe it's okay if they die.' No! People shouldn't die in avoidable accidents! And you stick to that theme. And you hammer it home with everything you say. And the story backs that up. It's an authoritative tone from beginning to end that says here's what we're trying to say. With authority, you know it when you've got it. It's something where there are really clear sides to it.

"The second is balance, which is extremely important for presenting something in context for public discussion. The voice of authority comes by being fair in the way the story gets framed in the newspaper. We're coming out very strongly that this is something the general public needs to know and look at and then they can decide to do what they want. That [report] better be fair. The essence of investigative reporting is truth.

"The third thing is a sense of outrage. You can be personally outraged when you write something, but you have to put [it] in context. Try to bring a little clarity to the situation, a little more truth. That's why you're an independent thinker on any particular topic you're writing about."

Q: How and where do you draw the line between advocacy and environment reporting?

A: "Writing with authority, it's fun to do. And it's challenging. And it's hard. People in the beach building industry don't like me very much because they think I'm against beach building. They think I'm against it as an issue personally. And I'm not. I see it as one of the very valid ways that we can protect priorities and also provide recreation for tourist dollars.

"I've been asked more than once – why are you so against us? And I'm not! I just am reporting things – true things – that they wish wouldn't come out. It's a spin game. And you gotta figure out – you don't necessarily come down in the middle every time. You gotta prove what you say. Check and double check and triple check the facts. The beach story is the one that probably gives me the most to stomach. The beach industry – this good old boy closed right up to Congress. They [the company hired to build the beach] had special things, they limited competition, they are constantly getting charged by the Justice Department with bid rigging. There's a long history. There are only two or three firms that do this kind of work, they limit foreign competition, and jump back and forth between lobbyists and congressional candidates, all in the same industry. And it just reeks!

"But I still gotta think to myself, am I being fair in the way we frame it up? As long as it's fair in the way it's framed, in a way that the average reader can look at it, it's fair. That's why journalism isn't truly objective. Too many judgment calls all based on the subjective learning that we've had from childhood on."

Q: How and where do you draw the line between advocacy and investigative reporting?

A: "There's no such thing as true objectivity, but there is such a thing as balance. Advocacy consists of one-sided propaganda from someone who is trying to further a single cause rather than general understanding of the topic. The public entrusts journalists to do the latter.

"There are two different types of advocacy. There's advocacy born out of self-motive, and advocacy on behalf of another person. Self-advocacy is what lobbying groups do. They say this is our cause and you need to think it's important. That's different than saying here's something where there's a waste of money and wrong things are happening to people, and that's not going to further your agenda.

"[However] if I'm asked to write a story about wife beaters or child rapists, my perspective is going to be that it's not a good thing. I'm not going to put someone in the story who rapes children, and interview him in prison and use a quote of him saying it's really what we all need to be doing to relieve tension. I'm not going to do that because there are moral imperatives that come before journalistic goals. My duty is to stop these things from happening. Your duty as a citizen outweighs your oath as a journalist in most cases. And even in all those cases, there's a creative solution so that you can do both. You can act the proper way journalistically, and you can also do your duty as a citizen when you get in a pinch like that, [such as publishing a column reflecting contact with someone who appears to be a suspect in a crime without going to the police directly with the information.]

"It's very easy to confuse investigative reporting with advocacy. Voice, framing, and point-of-view are the factors. There's not a problem with writing a story and having, as part of that story, some alternatives. But when you go much beyond that, then you're advocating something, you're moving away from being an independent. You move toward molding your community rather than mirroring it. When you put out a story that shows the police chief's a crook, you're going to change your community because the chief is going to get fired. That's not advocating that the chief gets fired, though. That's what an editorial would do. With a story, you have to balance it, and stop there.

"Mainstream weeklies or dailies are dedicated to both sides of the truth. If an article points to a problem and says this is the solution, period, that's being an advocate. That's imposing the might of the newspaper on the community, and that is advocacy. Let's flip the coin, though. What if we're writing a story about a family that is in real need of a liver transplant for their 2-year-old daughter and they're broke. What if we write a story that brings in $2 million worth of contributions and everyone lives happily ever after. Didn't we advocate on behalf of the little girl? Yeah. We did. And that's ethical. We decide. There are many more of those people in the world than we in the media write about. So it's an independent decision to write the story. Its not like the United Way came to us, told us about a person, told us to put it in the newspaper, and we said sure.

"There's a fundamental difference between what Amnesty International will write about itself and how I will use the organization in a news story about political prisoners of war."

Q: How do you do an investigative report?

A: "All my investigations start with a Lexis-Nexis search. Not only does it give me a knowledge base, because I trust what other papers write, but it shows me where the holes are in the topic. Then I go to the source chain.

"Investigative reporting at the *Sarasota Herald-Tribune* is individualistic, Bayles stressed, but it generally involves an editor regularly asking whether there's a story. That occurs at periodic intervals to help investigators assess what they have, where it appears to be leading, and what needs to be done to flush out the story. Bayles' style when developing an investigative report about the environment begins with finding out what everyone else has written on the subject in question.

"Usually, after I do my Lexis-Nexis search, I outline my larger stories because at that point I have a pretty good sense of where I'm going. I don't know what information is going to be in what particular paragraph. But I write a rough lede, where I'm going to go with this story. And this is what I use to sell my editors on something," said Bayles.

His outline is a five-part format for each story in the series, following in order: (1) introduce the concept in four or five paragraphs, maybe up to eight; (2) write the first subhead to lead into the breadth of the problem and what the report looks into, including between three and five examples demonstrating any national implications, and finishing by flushing out those examples and how they relate to Florida's problem (local angle); (3) write a second subhead to lead into the meat of the story – the statistics and arguments; (4) write a third subhead leading into a response from all sides; and (5) end with an anecdote, perhaps something whimsical.

"I'm very fond of a technique that *Newsweek* writers use a lot, where the end of it leaves you feeling like you just got hugged. I try to do that. I can't get away with that in daily reporting too much because it's almost a little preachy. But you can get away with it in your bigger pieces."

Q: What are some realities of the environment beat?

A: "Like most beats, there's never enough time to do all the stories you want to do. Often an editor will forget you're the environment reporter and assign you to a breaking news piece or have you fill in on another beat because many environment stories can wait. It's frustrating when you're in the middle of a 'weekender' (an evergreen that runs over the weekend – that doesn't have a day hook) on a local environment problem and you get pulled to cover a trial. But I guess that happens on any beat. I think it may happen more often on the environment beat because so many of the stories are left pressing."

Q: What does the future look like for environment reporting and for investigative reporting?

A: "At least for investigative reporters, I think a new responsibility will be the possibility of providing daily diaries online. Common thinking is that the Internet is going to change everything. It doesn't – it's just a different way to get information. You still have to double check it. It's just another way for people to put their views out there. Or their spins. Which is fine. But you still have to get the original document in your hands. You still have to talk to the mayor.

"I don't see any great changes ahead for the beat, or any instability about its future, but I do think there will be more reporting about global climate change and global disease like AIDS and SARS. That may bring about a sense among some of the elite environment reporters of a grander scale of projects and stories. But a very average mid-size newspaper like mine, it's going to carry on with a mix of daily journalism and longer pieces that bring things out.

"The environment is too big an issue in Florida to be ignored. So you won't see massive cuts in the environment in most papers in Florida. I can't think of a paper in Florida that doesn't have an environment reporter.

"I don't think environment stories are well suited for television news. I don't think you can do anything really well on TV. Maybe the news magazines can kind of get into some stuff. But the power of television often eclipses the power of newspaper in audience in mass and numbers. They're going to get the word out faster than we are, and more completely. But as far as in-depth journalism, TV is just not going to have the lasting impact of newspaper stories."

The *Sarasota Herald-Tribune* has a cable studio in the corner of the newsroom. Storied aired mostly tease to the newspaper version of a story, Bayles said. How to use the cable station for news is the hot topic of the moment at the paper. It is an ongoing debate in this multimedia newsroom, one of the first newspapers in the country to have a broadcast studio in the newsroom.

"How do you do a story on TV that you're also doing in print? For the most part right now, those reporters at the paper choosing to do TV as well as print use it as a vehicle to tease something they've written, and co-ordinate the tease with when the series or story is going to run. So the whole news report is a tease.

"This is exactly the stuff that's happening in our *New York Times*-owned newsroom. There is a discussion of how long can a TV piece be that backs up a newspaper report? We talked about two minutes, three minutes, and the producer said why don't we try five minutes? Why don't we try eight minutes? And everyone started nodding their heads. Newsmagazines keep a topic interesting for an hour. Why can't we keep it interesting for 10 minutes, especially if it's something as multi-faceted as [an environment story]," Bayles said.

[Questions also include whether to do a two-piece thing on the cable channel the day before and the day of a newspaper piece. Should the TV report be more substantial and contain different information? Also, there is sometimes major coordinating required with other broadcast outlets to get a story together. Sometimes the reporter is interviewed as an authority, Bayles said, speaking from experience.]

"That's the way the environment beat or any beat is going to change in the future. Reporters are going to need increasing flexibility. I don't think the content – what we're covering – is going to change. But the way we're doing it certainly will."

Tom Bayles' Reporting Process

Bayles described in detail his process for developing an investigative environment report that was nominated for a Pulitzer Prize. Referred as the "Beach Builders" series, it was sparked when a local man died in a dredging accident on the job. Bayles' investigation into the cause of death led to winks between special interests and law enforcement officials, and several unfavorable economic and environmental consequences of creating and maintaining sandy beaches all along Florida's coast to protect property values and maintain tourism.

Building up the beaches in Sarasota was aimed at preventing barrier islands in the Gulf of Mexico, which protect Florida's coastline from erosion, from disappearing into open water, Bayles said. Building a beach for Sarasota, he explained, involves dredging from barges in the Gulf of Mexico, hence his analogy in the series to a floating construction site during a storm. The story developed into a three-part series including exposes, explanations, and interpretations that took the concept of beach building to a new level of understanding for the public by illuminating government and politics, the environment, economics, and law enforcement and looking at them together, as a whole.

The newspaper used three voices in the report. In the first section, it spoke for people in a weak position. In the second section, it spoke for the elite or comfortable. In the third, it section spoke for wildlife. The newspaper's frame was to show injustice. Its point of view, its intent, was to say the injustices were avoidable, that they shouldn't happen. Period.

The Details

"So, one of our home boys died on a barge. And we get to thinking, let's figure out how he died," Bayles said. "It's a death, it's a young guy, it's an accident. Let's see if this is a preventable accident. So I make a few calls and all of a sudden within one day I learn there are probably seven more deaths in the Southeast United States. So the flag goes up. So then I started getting records and I realized that the vast majority of these were due to cost-saving measures on the dredging firm's behalf, that were overseen by the federal government, and that allowed safety measures to be missing," Bayles said, escalating further into outrage with each word, although Bayles describes that passion as nothing more than "professional interest."

Bayles filed piles of Freedom of Information Act requests and got some help from people in the government and the Army Corps of Engineers. "Most of my stuff came, as it always does, from people whom I made my friend and they just gave it to me. The FOIA [requests] asked for accident records in all 41 Army Corps districts that border beaches in America – Alaska and Hawaii, too. And as they came in, I learned how to streamline the FOIA system by convincing the FOIA officers that it would be best to just call me back with it rather than going through all he expense of copying it and mailing it. And I got a lot of Coast Guard records with lines blacked out. And a lot of OSHA records – they're online. They're pretty easy to do. I had a computer-assisted guy up my back (backing him up), so he helped a lot, too.

"In the beginning when we were trying to frame up what we were going to say, we found first of all right off the bat that these dredging companies, which were employed by the government, were not following the government's own safety rules and things like that. Some have rusty hand rails and the gears are supposed to be covered. A big gear wasn't and sucked one guy in. So I cross-referenced OSHA's databases with the records I was finding from these companies, and a guy came to me who was a disgruntled safety officer no longer working for anyone involved in dredging who had kept really good records for the past 15 years at one of the government contracting companies. So we had clerks type all that into a database.

"So at this point we know we have a story about this vaunted industry. Before we wrote what we did, no one ever wrote anything bad about beach building. It was always 'Save the community' and 'It's the greatest thing since sliced bread!' Nobody ever asked these questions. So we knew we had a story that we, as a paper, could say we knew these accidents were preventable. Now who is going to dispute that?"

At that point Bayles knew he had a "Gotcha!" story, the sort that doesn't come along very often. In the past, investigative reports often stopped there. But Bayles wanted to see what else might be wrong with beach building. "We wanted to see where we could effect very positive social change for people in a weak position," Bayles said. He began talking with dredgers and their families, learning what they go through on the barge and how many times a finger or foot or limb gets crushed on the job, discovering one company had averaged three injuries a day for 10 years. He continued to "drag string" for a story that would say everyone thinks beach building is so great but it may not be so great for the people who do it. That was the voice Balyes and his editors decided they'd use for the first day of the story.

The second day of what would become the three-part series took on a different voice. This one said what a great thing it is for the communities along the new two-mile stretch of beach being built – that it is only costing those communities $100,000, and the local politicians are great because they got the federal government to do it. But, Bayles said, "If you go in the back end of that figure, that $100,000 is the tip of the $55 million, 50-year cost of this beach. The federal government mandates through water resources [laws] that once they decide to make a beach sandy, they agree to keep it sandy for 50 years. So, every time a storm comes by and washes the beach away, they put it back. Sixty-five percent of the funding comes from the federal government in most cases. So, the wheat field farmer in Iowa is paying for Sarasota's new beach. It's not the cost from the taxpayer that's getting the most benefit.

"In the balanced story we ran, we had the success stories – the Miami Beaches. And in doing that, I came to find there were environmental factors that were not looked into," Bayles said.

In the last part of the report, Bayles began his investigation into environmental impacts of beach building based on the premise that if you take 1.2 million cubic yards of something and dump it at the shore, there's going to be some effect on wildlife there. The question was how much. Bayles went to the Army Corps of Engineers and asked to see an environmental impact statement, which is required by state and federal law before any development can occur.

"So I get one," he says, "and it's 500 pages long and you can't read it. So I ask for a few more. Well, they're all the same! How can a detailed environmental study of Wrightsville Beach, North Carolina be exactly the same as for Sarasota, Florida? So I went to various professors, some independent voices critical of [beach building], and they pointed me to some of their students' work. They were right in the loop." Bayles found the dumping caused dramatic impact to the bottom ecosystems along the shore. "The 500 little creatures that inhabit the beach get smothered, and there's damage to the fishes. We had the research department do a search of all the newspapers in America to find if anyone else had written about this stuff. I found a small newspaper that had reported on how a sea turtle got killed in the dredging of this one channel that needs to be kept deep for the military. The sand from the dredging is dumped on a nearby beach. So I called the people working that beach and asked how often that happens. And they started screaming at me! They said, 'It's not a beach nourishment!'

"And I said well, what is it?"

"Well, it's a channel dredging, so it doesn't count."

"I said well, where's the sand put? On the beach? What does it do to the beach?"

"It widens it."

"So what is it? I asked."

"It's channel dredging! It's channel dredging!"

"You know, well, bullshit. It's an artificial separation between the two. They're putting sand on the beach and in the process they're killing turtles. Well, I come to find out they get this special exemption. The private dredging industry is exempted by the federal government from its own endangered species regulations by allowing incidental takes – they call it a 'take' when it's really a death – of like 35 turtles a year. And one year they reached that number by April, so they upped it to 50. I mean, it's just ridiculous what people will do for beaches!

"So it has a huge environmental cost that no one ever reported about. Some people had touched on how much money it costs, and how committed it is, but everything else was never reported by anyone anywhere. Everyone always thought beach building was great," he said. The story cost $70,000 total to produce.

The reporting, Bayles explained, flushed out the voice the newspaper would use. "We had a good idea of what the voice would be after some initial phone calls, but the reporting told us what was wrong with this picture, and that was our voice. Men are dying while reshaping the beaches and reshaping the ecosystem, a lot of people are spending a lot of money to do this, and a lot of creatures are being killed in this quest in a futile attempt to keep a barrier island from moving."

Chapter 2: David Helvarg

An accessible nonconformist with a searing sense of humor and precise pen, David Helvarg is an independent multimedia journalist with an international reputation and more than 32 years of experience. An avid scuba diver and body surfer living on San Diego's beach in 1977, Helvarg began reporting about the environment that year with a story about the liability of mining the oceans for minerals.

Using his reporting background, the New Yorker by birth and historian by degree obtained a California private investigator's license early in his journalism career. In 1973, Helvarg was a freelance urban combat radio reporter in Northern Ireland covering the conflict between the British and the Irish Revolutionary Army. Ten years later, in 1983, he was deported from El Salvador while reporting for *The Associated Press* about a massacre of civilians.

During Helvarg's career in journalism, he became a television news and documentary producer concentrating on environmental issues relating to oceans. He later went on to create programs about the environment for *PBS*, *Discovery*, and other television venues. His viewpoint currently is heard on Public Radio International's program, *Marketplace*, and read in *Popular Science*, *The Nation,* and *Slate.com.* Helvarg has also led investigative reporting training sessions for the International Center for Journalists in the United States and Europe, and is the author of *The War against the Greens* (1994), and *Blue Frontier: Saving America's Living Seas* (2001).

Cited as one of the top environmental journalists of the 1990s by author Michael Frome in *Green Ink*, Helvarg's instincts have led him to track environmental issues by following the money and looking at how contending political and economic forces respond to changing science and policy. In 1988, Helvarg won an Emmy Award for his work with AIDS Lifeline, a televised awareness campaign. He also was awarded two National Association for Interpretation awards for Communications (1989, 1991), the Nike Earthwrite Award (1997), National Health Information Award, and a Golden Eagle Award (1999).

Today, Helvarg heads the Oceans Awareness Project, a nonprofit organization in Washington, D.C. he established in 2003 to work towards restoring the world's waterways through education and legislation. Helvarg is also managing editor of *Multinational Monitor,* a magazine founded by consumer advocate and former Green Party presidential candidate Ralph Nader.

Excerpt from Chapter 5:
Oil and Water, in Blue Frontier:
Dispatches from American's Ocean Wilderness

Published by the Sierra Club, 2006.
By David Helvarg

Amberjack is the ultimate Tinkertoy. An active drilling rig, it towers 272 feet from the waterline to the top of its bottle-shaped derrick. It's density of utilized space is a structural salute to human ingenuity. The rig has a four-story metal crew building, helipad, flare-off tower, tanks, processors, compressors, drill deck with 8300 feet of piping stacked 12 feet high, 1000 barrels of drilling mud, mud shakers, cement, two big yellow cranes, an office shack, lifeboats, and hundreds of other flow-pipes, tubes, racks, gears, lines, and computerized systems, hanging out over either end of its legs on wide, thick steel shelves. You know who ever designed this thing doesn't waste closet space at home. Still, from the air, Amberjack looks small and somewhat fragile set against the vast, white-capped expanse of the gulf's deep blue waters.

The winds are howling close to 40 knots today, the swells about 12 feet, and with an extra half million pounds of drilling gear on board, you can feel some sea movement on this platform. Once inside we are given a safety lecture and told to remove rings and Velcro watch bands to avoid "degloving injuries," where the skin and muscle can get ripped off your hands. We're introduced to Cary "Call me Buba" Kerlin, the red faced spherically shaped "Company Man."

"Might look a little dirty," he warns us. "We've been getting a lot of gumbo mud while we've been drilling." Gumbo is a heavy clay-thick, gray-black mud that's hard to wash off.

As a drilling supervisor or company man, Buba has been around the oil patch, having worked Colombia, California and Alaska as well as in the gulf. Before that he spent 12 years with the US Fish and Wildlife Service doing environmental assessments on oil company dredging canals, "till they made me an offer to get out of government and into industry and I became oil trash," he grins.

Under the company man is the tool pusher, or rig manager. Then there's the driller who controls the drilling console, the skilled roughnecks who work for him, and the less-skilled roustabouts or general assignment workers. There's the mud man, or fluids engineer, who runs the lubricating muds (polymers, clays, dirt and additives) that circulate down the pipe string. Several stories above them all stands the derrick man on his monkey board, a small catwalk from which he handles the high end of 42-foot sections of pipe. As the pipe tilts up towards him, he leans out almost horizontal in his harness to grab the top of the pipe and align it with the heavy rubber fill-up tool that adds drilling mud to the pipe string.

Buba takes me up to the drill deck. It's a noisy, thrilling scene; a choreographed dance of steel pipe, muscle, and machine. The cranes lift the pipe to the roughnecks and roustabouts in their hard hats and steel-toed boots who manhandle it into position below the derrick with its massive yellow top-drive and block. I stand near the console on the water-slick deck watching the crew working the hydraulic tongs around the pipe-stem and threading it into the hole with a creaky slow rotation before the top drive begins its work. Right now they're down to 8387 feet. With more than 36 other wells already down there it's a directional driller's nightmare. They're drilling this hole at a 45-degree angle, although they're capable of slant drilling like a boomerang, going down and then up again. One of the crew has a T- shirt that reads, “”New Rig, New People, New Records.”

Another 42-foot section of pipe is chain-winched onto the derrick floor like a skidder-pulled log coming up a clear-cut hillside. I move forward and begin taking pictures of the red-helmeted derrick man as he leans out from his monkey board like a trapeze artist to grab the 13 -3/8-inch pipe top and begins shifting it around to line it up with the rubber mud hose dropping down on him from above. I'm carefully lining up my shot, when one of the roughnecks sneaks up behind me and slaps my ribs, letting out an animal howl.

I turn around quizzically. He's grinning happily. “I can't believe you did that,” another guy semi-shouts to be heard. I didn't jump at the prank because I knew there were no howling predator animals lurking on this rig - other than these guys, of course. On the way back to the helicopter I spot the crane operator on a break, standing on the catwalk outside his cab, licking an ice cream cone and staring off into the Blue Frontier.

For more than half a century the leasing of offshore oil and gas has been one of the linchpins of U.S. oceans policy. In 1896, less than 40 years after the first rock oil derrick was drilled in Titusville, Pennsylvania, the first offshore drilling piers were built out from the newly established spiritualist center of Summerland, California.

Soon the more secular oil men were at war with each other, hiring armed gun thugs, sabotaging each others piers and racing to suck up as much oil as possible before the wells lost pressure and had to be abandoned. Abandoned wells and badly managed gushers soon led to widespread oil pollution and fouled beaches. "The whole face of the townsite is aslime with oil leakages," reported the *San Jose Mercury News* in 1901. The resort town of Santa Barbara just up the coast quickly moved to ban oil piers, fearing their impact on tourism and beach life. By the 1920s the State of California had made several feeble attempts at regulating offshore oil by charging a 5 percent royalty, but this low-cost legal structuring had the unintended effect of creating a rush of lease applications by oil companies tired of the wildcatting competition. As charges of corruption and evidence of pollution mounted the state legislature was forced to take stronger action, placing a moratorium on all new offshore lease sales in January of 1929. By then the Standard Oil Company had developed slant drilling technology, which allowed it to tap into state controlled "submerged lands" from its onshore rigs in Huntington Beach, California (a practice later ruled illegal).

Louisiana was going through a similar oil boom in its southern swamps, lakes, and marshes but with its thin coastal population and lack of recreational opportunities along much of the flood-prone Mississippi Delta, there was little resistance to the blow-outs, fires and other pollution taking place there. In fact, many of the area's settlers had always made their living through economic exploitation of the swamp from fishing and trapping, to market hunting of ducks, to old growth cypress logging which resulted in the near extinction of the trees just around the time the oil-companies arrived.

"Oil provided an alternative," writes University of

Southwestern Louisiana Professor Robert Gramling, "and the shift from one exploitative use of the region to another seemed natural and unproblematic." In the 1930s Gulf Oil and other companies, having developed drill barges for use in the swamps, began dredging hundreds of miles of canals to access their claims. These canals and associated erosion and subsidence became major contributors to Louisiana's subsequent land loss.

While it was becoming clear that onshore salt domes and other geological features associated with oil also extended offshore, the world refused to recognize national claims to economic resources beyond the 3-mile limit. But World War II changed all that.

As early as 1943 President Roosevelt, agreeing with Secretary of Interior Harold Ickes, wrote that the 3-mile limit, "should be superseded by a rule of common sense. For instance, the Gulf of Mexico is bounded on the south by Mexico and on the north by the United States. In parts of the gulf, shallow water extends very many miles offshore. It seems to me that the Mexican government should be entitled to drill for oil in the southern half of the gulf and we in the northern half of the Gulf. That would be far more sensible than allowing some European nation, for example, to come in there and drill." 8.

By the end of the war, seeing no naval powers likely to challenge U.S. claims, Ickes wrote up a proposal asserting the U.S. right to drill for oil on the submerged lands of the continental shelf.

The State Department strongly opposed this breaking of a legal precedent that went back to the Dutch lawyer Hugo Grotious's 1609 brief *Mare Liberum* which (with the backing of Queen Elizabeth and the British Navy) established the principle of open seas.

Still, Roosevelt's Secretary of State Edward Stettinius had bigger fish to fry with the upcoming Yalta conference, which would determine how post-war Europe was to be broken up. As a result, Ickes was able to get a less experienced State Department official to sign off on his plan. In March 1945 a gravely ill Roosevelt gave final approval to the Ickes plan. Following Roosevelt's death, the atomic bombings of Hiroshima and Nagasaki, and the unconditional surrender of Japan that marked the end of World War II, President Harry Truman announced America's claim to its Outer Continental Shelf oil on September 28, 1945. This became known as the Truman Proclamation.

The political power of big oil was already well known, from the break-up of the Standard Oil trust, to the Teapot Dome scandal, to what became the first scandal of the Truman administration, resulting in the resignation of Ickes in February, 1946. Ickes resigned to protest Truman's nomination of former Democratic Party treasurer Edwin Pauley as Undersecretary of the Navy. In that position Pauley would control U.S. Naval Oil Reserves. But Pauley was also known as a bag man for the oil companies and during the war allegedly told Ickes and Roosevelt that if the federal government didn't challenge state claims to offshore oil (where the companies felt they would get a better deal), the oil companies would make major contributions to the Democratic Party.

Ickes and other witnesses recounted Pauley continuing to push this idea on the train returning from Roosevelt's funeral. With the press hot on his trail, Pauley withdrew his name from nomination.

By the end of World War II, the Texas' and Oklahoma wildcatting oil-boom days were long gone. The post-war industry was consolidating its political and economic power, as graphically illustrated by the oil derricks pumping on the front lawn of the State Capitol in Oklahoma City. This left a small independent company founded by former Oklahoma Governor Robert S. Kerr and his friend Deane McGee with little chance of securing any top-grade land leases. So Kerr-McGee decided to gamble with some new technologies.

The end of the war had brought experienced navy men and surplus Landing Ships, or LSTs, home to America's coastlines. In 1946 the Magnolia Petroleum Company (later to become part of Mobil, and then Exxon-Mobil), used navy war veterans to begin a drilling operation in the gulf 5 miles off Louisiana on a platform made of wood and steel. Surplus navy ships housed the crew on the leeward side of a nearby island and shrimp boats transported them back and forth. Because it drilled a dry hole, history has tended to ignore Magnolia's breakthrough effort. Credit for starting the offshore industry usually goes to Kerr-McGee, which established a platform anchored by navy surplus ships 12 miles south of Terrebonne Parish in the fall of 1947. Two and a half weeks after it began drilling in 16-18 feet of water, a roustabout called his boss on shore on the morning of October 4th and told him oil was collecting in the drilling mud. "Well skim it off," his boss replied. "Skim it off. Hell. There's barrels of it," the oil worker declared, heralding in one of the great new frontier energy booms in American history.

Conflicts between coastal states and the federal government over who would get to claim royalties from this bonanza escalated in the courts and in the press until 1953, when Congress passed the Submerged Lands Act. This gave states control out to the 3-mile limit, and provided for federal jurisdiction over all Outer Continental Shelf (OCS) submerged lands beyond.

With the rules now firmly established investors and speculators headed offshore, including the son of an eastern Senator, veteran war pilot and future U.S. President named George Bush. Bush was among what Fortune magazine called the "swarm of young Ivy Leaguers," who descended on the "isolated west Texas oil town" of Midland in the early post-war years, anxious to make new fortunes apart from those they were already in line to inherit. Bush and his partners formed Zapata Oil, named after the Marlon Brando movie *Viva Zapata!* that was playing in a Midland movie theater at the time.

"Hugh Liedtke and I got in the offshore drilling contracting business and worked out a deal with LeTourneau and built three legged rigs back in the mid-50s," George Bush later recalled. "And they went on to become prototypes for drilling equipment that sat on the bottom. Our timing was good. We suffered a couple of setbacks. The first rig bent a leg or did something, had to be hauled into Galveston to be reworked. And the third or fourth one, the third one, disappeared in a hurricane, just vanished. I went out. I've never felt my eyeballs actually ache. I was flying in a single engine plane with Hoyt Taylor. We were looking for any sign of it. We'd taken the people off and it was gone. A $6 million dollar investment, that would be more like $76 million today. I loved the business. It was pioneering, the LeTourneau design, and also our people were pioneers. I felt we were in on the early stages of a marvelous business."

Offshore drilling would also prove a marvelous cash cow, generating over a trillion dollars for the oil industry and over $125 billion for the federal government during the second half of the twentieth century. Today, offshore oil and gas royalties and lease sales provide around six billion dollars a year in state and federal revenues. About a third of that goes to Louisiana, Texas, Alabama, California and Alaska. The remaining four billion makes up the U.S. treasury's second largest source of revenue after taxes (in close competition with Customs tariffs).

David Helvarg on Environment Reporting

Q: What drew you to environment reporting?

A: "I found myself increasingly drawn to hard-edged environmental stories. In large measure this was a case of finding an ecological niche that needed to be filled. I noticed that most political and investigative reporters tended to view the environment beat as either uninteresting or low-prestige, while many environment reporters came to the beat from a science or nature writing background. And, just curiosity kind of draws me to want to have the adventures, to do the first-hand reporting so that, you know, I'd rather, if I'm going to write about nuclear waste in the Pacific Ocean, I'd rather dive the site myself than see it in documents."

The perks of the beat are secondary for Helvarg. While gathering investigative reports about AIDS, immigration, and a security break down at a nuclear power plant, he had the opportunity to probe stories about high-seas driftnets, military dumping of toxic waste at sea, and the car bombing of an environmental activist.

Q: What personal lessons have you learned from environment reporting?

A: "I was surprised at the level of passion and commitment that many environment reporters, particularly those working outside the United States, brought to the beat. While still critical thinkers, they tended not to fall into the traps of cynicism or careerism that many other reporters have. It reminded me of the reasons I first got into journalism [which helped refresh his commitment to the craft]."

Q: What personal lessons have you gleaned from investigative reporting about the environment?

A: "We're at a critical point. A year ago [2002] I thought about returning to war reporting in Iraq or becoming an advocate for ocean protection, exploration, and restoration. In part I decided that while we'll probably always have wars, we may not always have wild fish."

Helvarg said too little investigative journalism is focused on the economic and political interests driving how people use and abuse critical ecosystems. As a result, while continuing to "follow the money," he also became an advocate within the field, encouraging colleagues in IRE and other journalism groups to see the environment as an opportunity for new kinds of storytelling.

Q: What makes the environment beat meaningful for you?

A: "The best reporting I do is from a first-person perspective. It's symbiotic if I can go on great dive adventures that also become the basis for radio reports, articles, TV segments, and books that help inform the public on the risky state of the world's coral reefs for example, and perhaps inspire them to take corrective action.

"For a time, my need for a larger social purpose was met in providing people with information and points of view they might not otherwise have access to. A deeper personal sense of satisfaction came when I started doing investigative reporting. It happened at a time when easy satisfactions were few and the inquiries I was making were putting lives – my own and others – at risk.

"At best, investigative reporting is fun. It's puzzle-solving. I won a local Emmy and generated some prison. [But he hates investigative reporting's monotony and repetition, he said.] I didn't know it would be tedious at times. It's probably good that I didn't. On the other hand, you're digging for hours in some really boring stuff and there's an 'Ah ha!' moment, and the pieces fall into place, and you have the factual metaphor you're searching for."

Q: How do you view investigative reporting?

A: "[Investigative reporting] is about finding the truth, where the truth is often up for grabs. It's often being, you know, information is a weapon in conflict. And the truth may not conform to one side or the other. You may find a lot of sleazy journalism – the kind of he said, she said reporting that you see more and more, particularly in broadcast journalism. Sleazy because neither side may be right. And the truth isn't always in the center, either.

"I was captivated by the Watergate break-in, which was my introduction to investigative reporting. But it was really underground news in the 60s and 70s and early 20th century renegade reporters who inspired me to become an investigative reporter. [Readings in college led him to] identify with the participatory journalism practiced by John Reed, Carleton Beals, and Upton Sinclair. I discovered a contemporary resonance in the boisterous, often insightful and occasionally libelous voice of the underground press. But it was Reed who I most wanted to emulate.

"There was a huge underground press I read avidly. It traveled an interesting line between exposé and libel: very outrageous, very challenging, and I think it was a training ground. I was a stringer for an underground press syndicate when I went to Northern Ireland in the early 1970s to cover urban guerillas there.

"Most investigative journalism lacks a kind of writing that connects to people's lives. You have to take investigative journalism skills and use them to tell stories and impart information in a way that has a personal and visceral connect to the reader."

Q: What criteria do you use for moving the public with your investigative reports?

A: "I think I got to be a good writer being immersed as a participant observer – first in Northern Ireland, and later when I moved to San Diego – where I edited a weekly muckraking investigative paper in a town that had a happy news reputation."

Helvarg said he learns something about himself from every story he writes. His reports read like personal experience stories. In an analysis of Helvarg's work for this research, his criteria for moving the public through investigative reporting seemed similar to the components that make a personal experience story significant: statistics, for instance, that help to illustrate the human condition and show how it is changed by circumstances and events – by history as it happens. Fiction techniques applied in journalism. Time elements for further relevance. Details. Details. And more details for context. No generalizations or clichés. Those components appear to be selected subjectively, depending on what he perceives as the substance of the report, and the nuances of the stories that prompted the investigation, as well as, in his words, "the report's impact on democracy and injustice."

Q: How and where do you draw the line between advocacy and environment reporting?

A: "Journalism is the roughed-out sketch of history, in which I have a personal stake. Voracity is important to me. I'll work harder than most reporters I know to get the facts right and my ducks in a row. As a freelancer, I don't have the deadline pressure generally; and if I do, I won't put stuff in that I haven't checked. It's an internal thing. I get very upset if I get something wrong.

"[Initially,] I thought [journalism] was advocacy for the powers that be. I thought it was essentially a reflection of the establishment. But a *New York Daily News* reporter at the Democratic National Convention in Chicago in 1968 changed my 17-year-old view when he took off his helmet and gave it to one of the girls protesting with us. The guy said to us, 'You know, I always respected the police and didn't like you kids calling them pigs, but the way these guys are acting today, they are pigs!' A few hours later we ran into him again and his head was bleeding. I think that was my first very positive impression of a reporter.

"[Journalism] gave me tremendous access. I realized I could cross lines that other people couldn't. I am motivated by social engagement. In Northern Ireland, I quickly came to see the limits of advocacy reporting when confronted by Hiram Johnson's old lesson that truth is the first casualty of war. I quickly learned to develop survival skills as an eyewitness and an investigator in order to produce accurate and analytical reports too often defined by the gun, the car-bomb, and the disinformation campaign. I learned that my sympathies didn't matter - that all sides were going to lie to me because information is a tool of war. That made me realize I could not be both activist and reporter at the same time.

"Activism is not the same thing as propaganda. In Northern Ireland, I learned very quickly that regardless of your sympathies you have to operate independently, otherwise you become a propagandist. You get manipulated. Or else you become worse than a propagandist: You become a dupe, being manipulated because you're not being smart. I quickly realized there may be two sides to a story but they can both be wrong, or lies.

"Many of my close friend died while we were reporting from war zones. One, my friend, John Hoglund, a *Frontline* reporter, said, 'There's no such thing as objectivity in journalism. The thing about it is that I'm not going to be a propagandist for anyone. If you do something right, I'll take your picture. If you do something wrong, I'll take your picture also.' We all have a point of view, but we can't - or we shouldn't - sheat with it. You have to be factual. Different types of media outlets - magazines vs. newspapers, television vs. blogging - have different requirements and also different standards [for including point of view in a story].

"People sometimes perceive me as an advocate in the reporting I do. And when I do training overseas in places like Poland, it's very funny because these Polish reporters who came up under communism had to figure out how to get the word out in [their new] free-swinging atmosphere. At one point we were talking about how you have to get both points of view and [one of the reporters said,] 'Well, I'm interviewing this guy and he has one point of view, and I have an opposite point of view, so that's two points of view!' So you get a strong sense of advocacy where the issues are freedom and democracy.

"The [right-wing] American Enterprise Institute, *Washington Times* newspaper, and assorted members of the wise-use movement accused me of being a shill for the Sierra Club when War Against the Greens came out [on Sierra Club Books]. They were unable to find factual errors in the book, so they accused me of being a liar.

"I was threatened with libel twice, both on stories about the environment. The first came from a dolphin trainer who lost his security clearance because he talked to me without permission. The second was in 1984 when I reviewed the movie "Salvador." The main character got a libel lawyer in San Francisco to say I portrayed him as a drunk and a loser. I did. And I suggested to the publisher that there's no need for a retraction, that I could get half the press corps to verify. More importantly, the lawyer hadn't seen the movie. Neither lawsuit materialized.

"There are differences between environment reporters and other investigative reporters. There was this self-flagellation I found with environment reporters that I never found with other investigative reporters looking at political corruption or national defense issues and AIDS. In the last 30 years environmentalism has become a social ethic. Considering yourself an environmentalist and reporting on the environment - if you're a good journalist [i.e., one who doesn't slant facts and other information] - it is no different than a national defense reporter considering himself or herself a patriot. We have shared values as a society and environmental protection has been one of them.

Q: What decisions do you make when drawing the line between advocacy and investigative reporting?

A: "At this point in my career, I have decided to declare victory and move on. It's hard to stop thinking of myself as a journalist. But I do see that in setting up a non-profit advocacy group my status changes. It's a return to activism in a way, 30 years later.

"The one ocean resource not being fully exploited is good investigative reporting on waste, fraud, and abuse concerning salt water special interest operations and the agencies that are supposed to regulate them.

"I consider both my books my best investigative work, but Blue Frontier comes from the heart, while War Against the Greens is more of an intellectual pursuit.

Author's note: Both books have strong viewpoints, a requirement for making book-length journalism a good read.

Early on Helvarg noticed some differences between environment reporters and other investigative reporters. "There was this self-flagellation I found with environment reporters that I never found with other investigative reporters looking at political corruption or national defense issues and AIDS. In the last 30 years environmentalism has become a societal ethic. Considering yourself an environmentalist and reporting on the environment – if you're a good journalist [i.e., one who doesn't slant facts and other information] – is no different than a national defense reporter considering himself or herself a patriot. We have shared values as a society and environmental protection has been one of them."

Q: How and where do you draw the line between advocacy and investigative reporting?

A: With Helvarg committing the rest of his career as a writer of advocacy to heightening awareness about oceans, part of his function will be to press newsrooms to add a specific beat covering oceans.

"The one ocean resource not being fully exploited is good investigative reporting on waste, fraud, abuse concerning salt water special interest operations and the agencies that are supposed to regulate them.

"It's hard to stop thinking of myself as a journalist. But I do see that in setting up a non-profit advocacy group my status changes. It's a return to activism in a way, 30 years later.

Helvarg considers both his books his best investigative work, but adds *Blue Frontier* also comes from the heart whereas War Against the Greens is more of an intellectual pursuit. Both have strong viewpoints, a requirement for making book-length journalism a good read, Helvarg said. The earliest example of a call to action in *Blue Frontier* comes when Helvarg writes, "This is why there is a desperate need to develop and expand not only our biological knowledge of the seas, but also an active and educated political constituency to protect the oceans' living resources. Unfortunately, our politicians and national leaders seem to be suffering anoxia of the brain when it comes to understanding the value of our living Blue Frontier" (Helvarg 2001, p. 5). *War Against the Greens* relies strictly on description, dialog, statistics, facts, setting, characterization, plot, narration and detail to carry the author's point of view until the last chapter when Helvarg directly states his conclusion:

If Wise Use's attempt to sponsor a 'holy war against

the new pagans' of environmental regulation and reform is fully exposed to the public – the right-wing terrorism and vigilante violence, personal profiteering, political sabotage, dirty tricks, and disinformation – the public's reaction will almost certainly force the transnational's into abandoning Wise Use. (p. 459)

Q: What does the future look like for environment reporting and for investigative reporting?

A: "Conditions allowing the ability of ecosystems to sustain themselves in the face of growing environmental abuses will keep the environment beat alive for a long time to come. There's a collapse of the natural systems that sustain us, so look for more investigative reporting in this area.

"I used to say that it's better to come to reporting out of something you're passionate about and want to write about than it is to just decide you're going to be a journalist and study journalism. I used to say *that* when people came out of journalism school. I have no training in journalism other than on-the-job. I think on-the-job works best. Today, journalism students don't *learn* journalism, they *go into it*.

"I came to reporting out of my interest in history and activism. So it was following my interest of how history impacts all of us, which I learned early on from my parents' experiences. I was attracted initially to war reporting and found that you have to be more investigative because you aren't getting the truth from either side."

David Helvarg's Reporting Process

While teaching the basics of investigative reporting, Helvarg said that he stresses doing background searches of news clippings and other materials, tracking original documents, doing initial interviews and follow-up interviews, developing varied sources, getting all sides of the story, keeping organized notes and tapes, verifying all information, and being persistent. Working as a private investigator gave him additional tools and databases and places to look as well as keeping more complete notes and transcripts, detailing times and places, and doing day-to-day searches of court records, legal data, and property. It gave him experience and kept him practiced, he said.

Helvarg follows 10 general rules when producing investigative reports. They are:

(1) Study your interviewee in advance.

(2) Go in prepared, with notebook, pen or pencil, and tape recorder.

(3) Make a list of five or six questions. Others will come to you.

(4) Observe etiquette. Get off to a good beginning.

(5) Establish eye contact and stick to the subject.

(6) Ease off before leaving. Close with an open-ended question like, "Do
you think we've covered everything?"

(7) Keep the door open for further communication.

(8) Transcribe your tape while you can still decipher your notes.

(9) Save your notes.

(10) Learn from the Freedom of Information Act.

Chapter 3: Tom Meersman

For more than 25 years, Tom Meersman has reported about pollution, national resources, conservation, and energy issues in the Upper Midwest. He is currently working for Minnesota's largest newspaper, the *Star Tribune*, based in Minneapolis. Before joining the paper, he did the same kind of reporting for Minnesota Public Radio. His reports include deep looks at air and water quality, nuclear power, forestry, agricultural feedlots, garbage and recycling, wind energy, motorized recreation, deformed frogs, state and national parks, Lake Superior, the Mississippi River, other lakes and streams, and endangered species.

A native of Spokane, Washington with Belgian ancestry, Meersman's interest in U.S.-Canadian relationships relating to the environment led him to do his first environment reports. Growing up, his first exposure to investigative reporting came through CBS television's "60 Minutes," which he watched for years.

Meersman's career is marked by numerous awards. He holds both bachelor and master degrees in English and a teaching certificate for secondary education. Before becoming a reporter he taught high school, worked as a live-in counselor in group homes for emotionally disturbed adolescents in San Francisco and Madison, Wisconsin, and was a visual arts administrator for the Metropolitan Cultural Arts Center in Minneapolis.

Over the years, he said he has learned how to distinguish between "junk science" and the legitimate, peer-reviewed kind. Unlike many reporters, Meersman is comfortable with numbers. These "numbers" include how health risk assessments and computer modeling can be used as factors in determining environmental policy. His familiarity with the history of several advocacy groups and trade associations, their leadership, constituencies, and major concerns helps, too, he said.

Meersman is among the most highly respected newspaper reporters covering the environment today. What gives him an edge, he said, is knowing how things work: the makeup, authorities, jurisdictions and procedures used in Federal, tribal, state and local agencies that deal with health, environmental protection, and parks. Knowing the basic tenets of environmental law as well as how to use the Freedom of Information Act, Minnesota Data Practices Act, the state's open meeting law and ex parte communications rules also help, he said.

Meersman's extensive service to the journalism community focuses on providing training in covering the environment for local and national reporters, as well as those from former Communist countries and elsewhere in the world. He is also a founding board member of the Society of Environmental Journalists.

Invaded waters: No one wins in a crash with a fish

Published June 15, 2004 in the *Star Tribune*
By Tom Meersman

Marcy Poplett jumped at the chance to take her personal watercraft for a short run on the Illinois River on a sunny afternoon in October.

It turned out to be a near-death experience.

After driving a few miles, she idled the craft near a bridge and looked at the fall colors.

"Every leaf was just gorgeous," she said. "So I'm sitting there and all of a sudden this big fish flops out of the river literally and hits me right between the eyes," Poplett said. "I'm not kidding. It knocked me completely out."

Poplett was whacked by a silver carp, an import from Asia that moved into the Peoria area about five years ago. The carp have a tendency to shoot out of the water when disturbed by passing motorboats. Weighing 10 pounds or more, they pack quite a wallop.

Poplett quickly revived, but found herself floating face down in the river, bleeding profusely. She saw her watercraft floating away in the current, heading toward a towboat that was blasting its horns. She passed out again, but a nearby boater, alerted by the warning blasts, came to her rescue. Poplett suffered a broken nose, concussion, black eye, injured back and a broken foot. She has recovered from her injuries and expects to be back on the river this summer - but never again without a companion.

Other boaters along the Illinois, Missouri and Mississippi rivers have reported dislocated jaws, facial cuts, broken ribs and serious bruises. Hundreds have been startled as the thin-skinned carp shot into their boats and flew to pieces as they hit seats, coolers, fishing equipment and depth finders.

"They shatter when they hit something hard," said Duane Chapman, fisheries biologist for the U.S. Geological Survey in Columbia, Mo. "They just get blood everywhere and they're slimy. If you have a pretty boat, it's going to be a mess."

Chapman, who is 6-feet-6, said one carp jumped high enough last fall on the Missouri River to hit him on the cheek while he was standing in a nearly idle boat. He said it was like "being hit by a bowling ball." He wasn't seriously injured.

Vivian Nichols of Hartsburg, Mo., is another fish-in-the-face victim. Last summer, she and her husband took a friend in their boat to see the jumping carp on the Missouri. As she idled the motor, fish started flying on both sides of the boat and it began to seem dangerous. As Nichols took her eyes off the fish to steer the boat away from them, a carp whacked her on the nose and broke it.

"The riverways aren't safe out there," she said.

Nichols and her husband, an occasional commercial fisherman, have outfitted their 20-foot jon boat with protective nets to lessen their chances of injury.

Some people are staying off the water. Steve Kelly, owner of American Sport, a hunting and fishing equipment shop in Havana, Ill., said that most of his customers are unhappy about the fish. "A lot of the women won't go on the river anymore," Kelly said.

The silver carp and its non-leaping cousin, the bighead carp, can grow to more than 50 pounds. They have exploded in portions of the Mississippi and its tributaries since they escaped from southern fish farms in the 1980s. "We're probably talking millions of fish in an 80-mile stretch of river here in central Illinois," said Mark Pegg, director of a field station for the Illinois Natural History Survey. "And they're heading north."

In Minnesota, a commercial fisherman caught a bighead last October in Lake Pepin, downriver from the Twin Cities. No leaping silver carp have been seen in Minnesota.

Upstream, Lake Michigan is threatened by a massive Asian carp invasion, which could spread to all the Great Lakes. Fishery officials in Illinois are scrambling to halt their progress by building a $7 million electric barrier on a canal 28 miles southwest of Chicago.

Fish farmers in Arkansas and Mississippi imported bighead and silver carp from China in the 1970s to eat aquatic plants and algae that grow in huge catfish ponds.

Now, the carp that helped fish farming have damaged commercial fishing. Dick White has been fishing near Havana for 25 years, mainly for bigmouth buffalo, a fish native to the area that competes with carp for habitat. He sells buffalo to nearby restaurants and said the Asian carp have nearly ruined his business.

"They're just thick," he said. "You put a net out and you'll get 500 pounds of them and you can't raise it out of the water."

Asian carp damage fishing equipment and so far have almost no commercial value, White said.

"I'm mad, I guess," he said. "Somebody turned them loose and they got into here. I don't think you can make somebody pay, but it's just a bad thing that this has happened."

Tom Meersman on Environmental Reporting

Q: What drew you to environment reporting?

A: Before Meersman became an environment reporter, he perceived the beat as a lot of news about pollution and parks and forests.

"There seemed to be a lot of stories about air quality and water quality [that didn't emphasize one argument over another]. I felt [the stories] had an impact on people, and on health, and on our general well-being. That there were a lot of problems and they weren't necessarily getting resolved.

"It didn't seem like a lot of reporting was identifying those problems. It seemed to me to be important to our welfare to pay attention to some of those problems. Not just for people's welfare, but for ecology and wildlife. It seemed like it would be interesting because there seemed to be a lot of problems that needed attention."

Q: What personal lessons have you learned from environment reporting?

A: "People rely on me to be smart enough and creative enough and attentive enough to come up with good stories and news stories and interesting stories. [The environment beat] is what we make it. I'm kind of an independent guy. I liked teaching because it was relatively independent. You make it your classroom, you control the class, you have a curriculum you follow, but generally you have a fair amount of freedom in how you approach things and teach kids.

"I feel that way about my situation at this paper, and previously at public radio. It's not top-down. People rely on me to be smart enough and creative enough and attentive enough to come up with good stories and news stories and interesting stories."

Q: What personal lessons have you gleaned from investigative reporting about the environment?

A: "Every word needs to be right. Those who are unhappy with the report could seize on any shortcoming or error no matter how small to try to discredit the entire effort. If there's anything about it that's not solid or absolutely right-on, if you are raising issues that maybe some people don't want to hear about, there may be someone who tries to discredit the overall report.

"I did write a report one time where one small thing was not correct. It wasn't a project, line-by-line fact-checking, but one small thing was not correct. The person who was the focus of the report came to the paper months later and wanted the paper to retract the entire story based on this one minor error. He had served on some panels that helped distribute funding for researchers. I had written something like that he served on boards that help distribute funds for research.

"He looked at that as a major flaw and was trying to discredit the entire piece which was raising a whole series of questions about his activities and his ethics that the article had laid out. People always say to fact check, fact check, fact check. but there are people who will come in and say the reporter turned out a bad piece of journalism if it isn't absolutely perfect."

Q: What makes the environment beat meaningful for you?

A: "I have this strong emotional and physical connection with the next generation that gives me a stronger feeling about passing on [a healthy and thriving planet]. Maybe it's because I'm a parent, and also have taught younger people. I tend to feel very strongly about keeping the quality of the environment *as good* if not improving it. It is absolutely important because you have to behave in a responsible way. Those are my values. You don't take an action that will spoil things for the future. There needs to be respect.

"Ultimately, it comes from that you're gifted to be alive. You're even more gifted to be in a country like this. Because we are gifted doesn't give us the right to despoil the good natural resource base that we inherited from other people. We don't have the right to wreck it.

"I don't trace this to my background, but I can certainly relate to it: In this part of the country, there are a number of native American reservations. Their attitude is to behave and live for the seventh generation ahead of you. That would have played absolutely *no* role in my values as I was growing up, but I think that philosophy reinforces what I already believe. That overall spiritualism – that there's a reason for you, and that you're living in whatever period you do and have to behave this way; but there's also responsibility that the people who came before you and especially the people who come after you keep the world in reasonably good shape.

"You can do stories that educate people about nature. [Meersman talked about a story he wrote about lynx coming back to Minnesota.] They have 30 confirmed, based on DNA analysis, and there seems to be a breeding population. A lot of them are in the Superior National Forest. And there are some theories about why they are coming back."

Q: What makes investigative reporting meaningful for you?

A: "Investigative reporting gives me the satisfaction that I'm doing what journalists are supposed to do, which is be critical and thoughtful, pick important questions and project them and educate people about them. With Investigative reporting, you get something extra special, which is shelf life. People will keep referring to it. Two years later people will ask for a copy. It has lasting significance. People will think about it and remember it much more than a regular story. You hope that will happen.

"If you do a strong investigative piece, it will have legs, so-to-speak. It will not be a story that people will read, say that's interesting, and that sort of goes away. In my case, investigative journalism, where you really take a look at something, has made significant changes in terms of bringing about discussion and changing policies. So, occasionally, your work can actually change things in a way that you can see, in addition to educating."

Q: How do you view investigative reporting?

A: "My perception was [investigative reporting] would involve a fair amount of time, collection of considerable documents, that you have to be organized enough to be able to categorize those documents and be able to get at them over a period of time. You'd have to have yourself together and know what you have, and be able to organize it and focus it.

"I knew that before I did any investigative reporting because I had seen other investigative reporters' files and boxes that they'd have for certain things. I knew that you'd have to collect a lot of stuff and that it would probably take a while, that you might need to use the Freedom of Information Act or other tools to get some of the information. There's a certain amount of discovery with investigative reporting. You collect as much information as you can to try to figure out what the facts are.

"Much good reporting is investigative. When I think of something as investigative, it's doing a lot more interviews and spending a lot more weeks pretty solidly on a story, even though it is not a six-page project or a series that takes three months or six months or a year to do. I like to think investigative reporting is not something that is only done by two or three people on staff. Many reporters who want to look at something seriously will do a circumscribed but nonetheless investigative look at something, within a certain scope, in order to write a good, solid story that the paper would be proud of and someday run on 1-A.

"I want to make a distinction here. I've always been interested in journalism that gets at things that are not obvious. I think of that as *enterprise journalism.* Spending just two weeks looking at something – I consider that to be enterprise reporting [even if] it involves getting documents and doing quite a few more interviews than you might normally do for a story, and might be a bigger picture look at something than just a narrow set of facts. That's a form of investigative reporting.

"Obviously there are some things that take much longer to do, and you have to have clearance in order to spend a large amount of time on a project. So those [projects] are clearly investigative, where you are off on a project and unavailable for most other assignments. In our shop, if something is a project it's an investigative piece of journalism. And in some cases reporters are physically removed from where they normally work to go work with a project team in a particular area of the newsroom. And they are generally off-limits for [other] assignments. It takes two or three months to do a project.

"I do a lot of reports that are not three or four month investigative projects but are enterprise stories, where you take something that is interesting and you do something with it. So I consider enterprise reporting to be investigation over a very short time span. But it's not line-by-line, word-by-word editing like a project that might involve six to 10 people weighing in at different stages. But it [takes] close editing like any story would. Also, we may or may not take it to a lawyer. We [wouldn't] do a lot of the other things we do with some investigative reports, like have the option of carefully packaging it with details graphics and lots of photos. If it's a two-week story you can do some of that, but you don't have the luxury of time to consider alternative packages and designs.

"Investigative reporting is not an either-or. I see a whole range of reporting where you can use investigation in a very limited way. You scope it right away, you're not searching through tons of stuff, you're basically trying to set a good story in context, and to broaden it, and make it more significant to people instead of This is Just One Place, one area, These People Have a Problem. Too Bad for Them.

"Investigative reporting might [locate] 10 similar situations and look at them all and maybe find out that the people responsible have disappeared and there's no money in the kitty anymore because..." Meersman said, letting the words hang. "Investigative reporting is when it's all part of the same continuity. Some things require a good, hard look separately. An awful lot of environment reporting is investigative. If you see the promise, and look to what the story is and develop it thoroughly without over-reporting or spending too much time, there's a lot of potential there for stories to be much broader than they were without having to be major investigations that come up with some big 'Gotcha!'

"Investigative reporting is thorough and inclusive, and it doesn't take cheap shots, and it doesn't go with unsubstantiated allegations. It's the opposite," Meersman said, alluding to traditional muckraking reports. With investigative reporting, "You are working your head off to get as much documentation as you can get from a full variety of sources. Muckraking, to me, also has the connotation of being quick turnover, not necessarily something that is deliberate and carefully done. I have this sense of muckraking as kind of superficial. It's easy to take shots at people in reporting that doesn't get nearly as deep as it should. You don't just put somebody on the spot or publicly embarrass them.

"On the inside, investigative reporting takes its toll on the reporter as well as everyone else in the reporter's life because of the emotional stress involved. You have to prepare other people in your life when you're working on a project or enterprise report and say "I'm not going to be around much in the next couple of weeks.' Or 'I'm going to be missing a lot of dinners." Inevitably there comes a crunch time on a project to get everything perfect. It's 16 hours a day, and it's not pleasant, because you're working toward something you believe in. People understand it, then, when you're grumpy and cranky because you're not getting enough sleep. You have to give up your other life for a while.

"One of the things that I may find in investigative environment projects is that the reporter writes with more authority. You don't have a preconceived notion of the story, you investigate, you let the facts lay themselves out for you, and that's what the story is. You don't try to spin them like somebody may in what I think of as advocacy journalism. On the other hand, when you present facts, by doing so you write with authority about something and sometimes that's not a topic or those findings are not something that anybody else has assembled. So you don't need to just call up somebody and get their opinion about something and contrast it with somebody else's opinion and say that b-c maybe nobody has an opinion about the new facts that you have discovered."

As an example, Meersman talked about a story about all-terrain vehicles he put together. "Nobody knew how much damage was being created by the vehicles on public land. We went out and viewed a significant number of those lands and wrote with authority about the extent of the damage. Some people might call that advocacy journalism because you don't have any authority other than yourself to make those conclusions. But you become the authority by doing the leg work and presenting those conclusions to the public.

"That is investigative reporting in general. Some people might interpret that as you taking sides and you being an advocate, but in fact you are testing the reality to find out what's true. You have these occasions where there's an open question out there that everyone claims to have an opinion about, but nobody really has any defined answers to.

"I do a lot of enterprise stories. I've always been attracted to stories. This is what actually got me into journalism. I saw public radio as an alternative to other broadcasting. I felt it was different. It was freer. You could get into things deeper, you could be more playful with your issues. It was interesting to me, and it's what got me into journalism. One of the things that drew me there [to Minnesota Public Radio] was that you could do [reports] that you wouldn't normally see elsewhere. I've always felt that a mission of journalism is to break new ground, to raise issues that are not being raised elsewhere, to shine a light on something that hasn't had it before.

"That's always been an important thing to me in journalism. It connects my previous life as an educator with my life now. I feel that I'm still a teacher as a journalist, and feel that I educate a big audience on specific issues. I'm essentially a teacher. To be a good teacher, I'm not rewriting press releases or reporting things that are already being taken care of. I need to take a look at issues in terms of the bigger questions, sometimes, and sometimes in narrow slices. I like to spend more time looking at something that any reporter might be able to file a story about on deadline. Sometimes there's a lot more to it.

"I might do a follow-up with concerned citizens, whether they be whistleblowers or regular people who [say something is being ignored] and, 'Isn't this an issue of some concern to you?' For example, a reporter writing about an oil pipeline leak (which happens pretty frequently here), the reporter, instead of just writing six stories over two years about pipeline malfunction should be able to discern that they're all coming from the same company and that there might be something more going on there with regard to that company's performance or its inspections or its infrastructure or its equipment, and whether it's upgraded or what. Instead of always taking [a story] in bit by bit, trying to realize what's going on and see the bigger issue, to be able to step back and do a full blown story on it, that's what attracts me to journalism.

"I've always been interested in educating about things that I think are important. A lot of that involves hard work, thinking, and reading, and paying attention to people who come to you with information, and making time to check up on those things.

Q: What criteria do you use for moving the public with your investigative reports?

A: "You don't need a degree from a prestigious university to do investigative reporting. You just need to be curious and care about what you're writing about. You can have the degree and not have those things, and you won't be good. The work is what you make it. That's true in spades in journalism.

"To move the public in journalism, let the facts lead to the conclusion instead of starting with a conclusion and trying to work things around it. News journalism is reporting, not persuasive writing. There is a lot of journalism that is persuasive writing – editorials and op-eds are journalism, but *news* journalism is reporting and not opinion or persuasive writing."

Q: How and where do you draw the line between advocacy and environment reporting?

A: "All life has an inalienable right to clean air, clean land, and clean water, but with that right comes a responsibility for us to help achieve that as a goal. In other words, I don't think people should have the right to demand clean air and then do whatever they want in their private life that would sabotage that goal. With he rights to those things, we also need to take responsibility to help achieve them, instead of just demanding it and not giving anything back."

That said, Meersman regards what has become known as *advocacy journalism* with some caution. "I view it with some caution because a reporter is playing a role in the story. I like documentaries, and if that's advocacy journalism, well, I like the idea of television in particular, like "Nova". I like the idea of an in-depth look at things. Some people might say that's advocacy by nature of the fact that you're picking a topic and doing a show on it. It doesn't necessarily [address] whether that show is balanced or not. Is doing a show on pesticides advocacy journalism? I'm in favor of advocacy journalism is that's what it is, simply by topic. I'm not in favor of advocacy journalism if it is unfair or unbalanced within the context of the report that's done.

"To be journalism, it has to be balanced. That's what journalists are supposed o do. Fair and balanced. That to me is legitimate journalism. Anything is fair game as far as topics. [If a report] is biased, it doesn't seem to me to be journalism."

"[Stories offering an answer to a problem can go too far.] First of all, I'm not sure it's the truth. Secondly, it may propose solutions that are not solutions at all. If there's a controversy or something that's hard to solve, if it was easy to solve just like that, it would have been done! So the problem I have with advocacy journalism is that it might be someone who lets their feeling get in the way of getting as close to the truth as possible, with the result that you mislead people and may actually propose solutions that are counterproductive.

"Some advocacy groups target a couple of issues while a regular environment reporter deals with dozens and dozens of issues. A reporter has a much broader set of responsibilities than just a single issue. Sometimes advocacy people call you and don't understand that. They are working on an issue 24/7. They don't necessarily have a perspective on the other important issues that environment reporters also need to pay attention to.

"The very fact that environmental issues get into the newspaper – and that some people don't want to see there – causes them to brand the journalist who wrote the story – to kill the messenger – saying you must have an agenda to be able to write about this in a way that is critical. For example, all terrain vehicle advocates (ATV) might have thought that their point of view was correct when it disagreed with the facts that I reported. Our facts showed there was widespread damage from ATV use, and the advocates said it was their opinion that only a few people were doing the damage, and that we were exaggerating how widespread it was. Our conclusion was based upon reporting and physically viewing and photographing damage in all four corners of the state."

Q: How and where do you draw the line between advocacy and investigative reporting?

A: "Both [investigative reporting and advocacy journalism] are the same in that they raise issues of legitimate concern. The purpose of raising issues is to get them into public discussion. They could lead to action. Different advocacy groups do their homework. Some produce some wonderful reports. For other groups, that's not true. But for journalists it has to be true all the time. You have to document what you do, and talk to a lot of different people. So there are similarities in terms of issues that are important. But the approach is different.

"The journalist doesn't take a side. You can come to conclusions and use research and facts to say his is a significant problem, you can somehow quantify that, but that doesn't mean you provide a solution. My job is to identify and document and portray and put out there an issue that should have public attention, then let people talk about it, and let other people decide what needs to be done. To me, it's enough.

"Advocacy is taking a bunch of data from EPA and putting one's own spin on it to say what you think the data says in two pages, and it doesn't necessarily say that at all."

Q: What are some realities of the environment beat?

A: "It's like this beat is new all the time," Meersman said. "The diversity of topics, and the people. The practical matter that you get daily stories, enterprise stories, and investigative stories. It's not like the crime beat where you have to do a story every day because there's crime or an investigative team that may only write three stories a year. There are a lot of topics, incredibly interesting people, scientists learning new things all the time.

"Some people have told me they got tired of the environment beat because they kept covering public meetings about opposition to landfills or NIMBY (Not In My Back Yard) this or that. But I think wherever you live there's always something pretty interesting." Meersman gave an example of a juicy story with wide appeal and local angles in many states, a story easily overlooked.

"There's a strong interest in Minnesota in the environment and the outdoors and nature," he said, providing an example where one can see the educator at work. "There was a story I did about a Superfund site in Ashland, Wisc. on the shore of Lake Superior earlier this year. We could have done that story in two days, just doing six interviews and saying this is a new Superfund site in Wisconsin and there aren't many new ones on the list. Here's the problem, here are the parties, here's what they say, blah, blah, blah.

"But we spent a couple of weeks on it and turned it into an enterprise story because the nature of the pollution was that it came from an old coal gas manufacturing plant. They used to make natural gas out of coal on the outskirts of many cities around the country. They would manufacture this gas and use it in the street lights before there was electricity. And because there were no pipelines until WWII, when natural gas really became an industry, our natural gas in a less pure form was manufactured on the outskirts of towns. So there are hundreds of towns in the Midwest that have areas of pollution from these old coal gasification plants.

"In this example, that's extremely problematic because a lot of the residue from this particular coal manufacturing plant in Ashland got into Lake Superior and is just beneath the sediment going down about a foot and covering about 40 acres. And they don't know how the hell they are supposed to get this contaminated sediment out of the lake. It's a multi-million dollar cleanup effort and they're debating how to do it.

"We could have done the story in 12 inches. But we ended up doing a front page story that jumped to the inside and that had a full page with it and it had maps of Wisconsin and Minnesota and showed the known sites of all these old coal gasification plants, and it provided a geographic context as well as a historical context that this legacy people 100 years ago left to us and we're having to clean up in some cases. Ashland is a good example because it's a very expensive cleanup because it involves Lake Superior. And there were all these lumber plants that were operating there at the turn of the century, so there were liability questions about how many of them, what did they use for creosote, how much came from coal gasification, how much came from the way they treated lumber.

"So this environment story turns into a historic story, just an interesting period story. We felt it was interesting enough to spend two weeks on it and have some nice maps. It was worth it."

Meersman said he considers that to be an investigative piece because "we got databases and looked up historically what is known, how many sites are known, how many have been cleaned up, how many are not cleaned up. We [focused] on this narrow window of pollution problems of the past and did a very thorough report. And a lot of people thought it was very interesting. Although the Superfund site was in Wisconsin, the company responsible for the cleanup had its headquarters in Minneapolis, which gave me a strong local tie."

Meersman has not had difficulty getting stories about the environment published. The position he landed at the *Star-Tribune* in Minneapolis was highly coveted, he said, going into some detail that will not be provided here. Suffice it to say, in Meersman's words, "It is a very competitive, sought after position."

"I came to the job with a background in environmental reporting, but my previous job at Minnesota Public Radio didn't require me to have any special background that qualified me to cover the environment. The *Star-*

Tribune did. They wanted an environment reporter who could hit the ground running. Initially, I covered the environment part-time for radio. Then it evolved into a full-time beat. Initially," he emphasizes, "because it was such a busy beat that it was difficult to define a chunk of time, and to convince editors that a particular story about the environment was so important, and to focus on it, and let other stuff go to cover it as best as we could. It was a pragmatic decision. As long as you can produce a story a day for our local audiences, we're happy if most of them can be environment. But it wasn't like a business decision, like them figuring out what listeners want, should we create an environment beat. it was fairly young in Public Radio's history when I started. As it grew bigger, it could afford to have people doing one or two beats each, instead of everybody doing general assignment. It wasn't a big decision that we have to cover the environment. They just wanted quality news. I was doing a lot of stories on the environment and it evolved into a full-time beat because there was just so much to write about. And they were interested in having good regional stories on the air."

Q: What does the future look like for environment reporting and for investigative reporting?

A: "I think there will always be investigative reporting. It's time consuming, but I think most newspapers realize it's part of their franchise. Newspapers have a special responsibility to have the longer, more involved, bigger sorts of things that get put before the public. It's part of newspapers' key mission, that they have extended, well documented issues explored as part of their regular mix. It's one of the hallmarks of newspapers and what I think they will continue to do.

"I think the future for environmental journalism is equally positive. The environment has matured as a subject over the past few decades. It may have been thought of just as parks and pollution in the pat, but now I think it's much more realized as the sustainability issue. Environment means air quality, which means something having to do with the way we design our roads and policies [concerning] fuels and vehicle emissions. It has to do with the way we produce our electricity and what our utilities have in terms of sources of fuel and power. It has to do with public health issues in terms of food safety and water safety. And it has to do with agricultural issues in terms of how much round water is used and who gets to use water anyway in areas where there's a water shortage. And it has to do with greenfields and brownfields and where we're going to be able to grow ultimately, and what seems to make sense, and what kind of pollution insurance is needed and how we deal with the legacy of the past when we are a society took some actions or allowed some behaviors that we now realize were not so good for the environment and need to be cleaned up.

"So environment is more a theme than a beat. It crosses many beats. There will always need to be a fair amount of reporting that is related to the environment, especially air and water quality, and transportation – whether you have an environment beat or not. It's part and parcel of all that's going on in the news about growth and development and how we change and what we get rid of and what we move toward, whether it be transportation systems or electric systems. The whole infrastructure of society has its roots in the environment. There are environmental considerations brought to bear on the choices we make.

"So the beat is safe, even if we don't call it a beat as such. It's a real important area that will continue to be there."

To be safe, Meersman has always been skeptical of his sources and kept them at arms length rather than forming friendships with them. Over time advocacy groups, he said, have become more sophisticated, and his skepticism of them "has matched their growth," he said.

"I think they raise good issues, but I see them for what they are: an interest group. Early on environment reporting would take what some of the environment groups had said and might not be as critical of what was backing up those kinds of claims.

"But the beat itself is more sophisticated now. Reporters know what questions to ask. [The beat] requires environment reporters to be a lot smarter to keep up with the science and the quantitative stuff and try to make some judgments: What does this mean? Is this a matter of concern?

"What we need to be more skeptical of is whether the government is doing its job and whether environment groups are playing it straighter than people were in the early days when we were looking at these gross physical problems in the '70s.

Tom Meersman's Reporting Process

"Before I get into a project, I try to write down a handful of questions that I want to answer, so that I don't learn too much and forget about what a regular person might want to know about this. And maybe some of the questions might be pertinent as I learn more. [This step] provides me a reality check to go back to. [It gives me a marker for saying] these are questions at that point that I thought a regular person would want to know in terms of information. You can't neglect these things. You can go deeper if you want to, but you have to get these things first because these are the things that would come to any reasonable person's mind when you bring up the topic.

"So when I've answered all those questions, and taken it further and found new things, a way I know I've got as much as I think I'll get is if in additional interviews I start to hear the same thing. [When that happens, I say] I think there's truth to this so I don't really need to go that much further. Its a hard thing to figure out if you've left some stone unturned. You get tired of it [interviewing] when you start hearing a lot of repetition. You just have to make a judgment call. If your editor is following the story very closely and hearing from you, sometimes they'll raise the question of whether it might be time to start writing."

Chapter 4: Charles Pekow

After talking with Charles Pekow, one wonders if institutionalization for psychological disorder is among the best grooming for prospective investigative reporters. From the experience, he gained a sense of being wronged, which gave him deep outrage at injustice and a strong motivation to right wrongs.

This sense of outrage possessed by investigative reporters is necessary, according to the literature and to Pekow as well as the other reporters interviewed for this effort. Now a long-time freelance investigative reporter and bicycle enthusiast in Washington, D.C., Pekow hails from Highland Park, Illinois, a Chicago suburb.

While involuntarily institutionalized by his parents for at least 10 months a year between the ages of 11 and 21 (1965 to 1975) at the Sonia Shankman Orthogenic School at the University of Chicago, Pekow was forbidden exposure to the outside world. He saw his first investigative report around age 16 on CBS-TV's "60 Minutes" news magazine while home for summer vacation.

Pekow's first taste of righting wrongs came when he protested to institution counselors that he was entitled to know what was going on in the world, and claiming that the denial of such a liberty was a violation of his civil rights. As a result, the rule was relaxed, and residents were allowed to watch "60 Minutes" on Sunday nights. The investigative reports intrigued, excited, and inspired Pekow, who shortly thereafter became hooked on a rebroadcast of Edward R. Murrow's "See it Now" series. "These were real inspirations," Pekow said of the stories. He also craved *Chicago Daily News* columnist Mike Royko's work and others "who would inspire me as a citizen to do something," he said.

Pekow, who holds a bachelor's degree in government from Georgetown University and a master's degree in journalism from Northwestern University, has no formal training in investigative techniques. Nonetheless, he has received several awards for his investigative reporting: one from the National Press Club, one from the American Newspaper Publishers Association, and three from Newsletter & Electronic Publishers Association. To date, he has gathered and written as many as 30 reports about the environment. By contrast, he has gathered and written several investigative reports.

General investigative reports take months, Pekow said. When the environment is involved, however, they take longer due to the complexity of the scientific jargon, he said. Several of Pekow's investigative stories have been suppressed, but not those concerning the environment, he said.

Lines of fire: The only thing standing between you and a deadly oil pipeline accident is Washington's most hapless regulatory agency

Published in *The Washington Monthly* Jan-Feb 2002
By Charles Pekow

Amanda Smith was enjoying a weekend of fishing and camping with her extended family along the banks of the Pecos River in New Mexico on Aug. 19, 2000. Early that morning, a few family members had set out to catch some fish, taking their lanterns with them to the river.

But instead of waking to the smell of fish frying, Amanda awoke to her children's screams. A corroded 50-year-old gas pipeline, which crossed the river next to the campground, had ruptured and sent a flaming fireball across the campground. Burning fuel rained down upon the campers. The fire was so hot it melted sand into glass and turned part of the concrete bridge into powder. Tents and sleeping bags turned to soup. The flames leapt 500 feet in the air, were visible from 30 miles away, and left a crater 86 feet long, 46 feet wide, and 20 feet deep.

Amanda Smith survived the fire only long enough to describe the horror of the scene to rescue workers. The explosion wiped out most of the Smith family, killing Amanda, her parents, her husband, her two kids, her brother and sister-in-law, their 22-month-old daughter and twin 6-month-old babies.

The New Mexico accident was just the latest in a string of horrific pipeline accidents over the past few years that have left hundreds of people dead or injured.

Corroded pipes, operator errors, and outside damage are taking their toll on the nation's aging pipeline infrastructure. The sprawling network of more than 2 million miles of pipe and 150,000 miles of underground tubes carries 20 trillion cubic feet of natural gas and about 14.4 billion barrels of hazardous liquids such as propane and petroleum each year, a figure likely to grow. The number of accidents is also on the rise.

The U.S. Office of Pipeline Safety (OPS) has counted 6,377 accidents between 1986 and August 2001. These incidents caused 376 deaths, 1,699 injuries, $1,140,697,582 in property damage, and a gross loss of 2,777,205 barrels of various oil fuels. (OPS can't quantify the loss of natural gas.)

Because of these accidents and the growing potential for new ones, safety experts and state officials have been pleading with OPS for years to mandate the kind of pipeline inspections that would have prevented the explosion in New Mexico. But OPS, a little-known agency within the U.S. Department of Transportation (DOT), has routinely rebuffed suggestions that would increase enforcement of existing safety rules and dragged its feet on implementing new ones. More troubling, OPS has shown an unhealthy relationship with the industry that it regulates, which, incidentally, provides all of its budget and has a huge financial incentive to avoid regulation.

As a result, the existing regulatory structure leaks like a sieve, with virtually no government interference and few sanctions for those who violate the rules, even when they cause significant environmental damage. OPS had cited the owner of the pipeline that killed Amanda Smith and her family, the El Paso Energy Pipeline Group, for safety problems with its pipeline in Arizona in 1997, and again in 1999. But it took the deaths of 12 people before OPS really jumped into action. The agency is now seeking a $2.52 million civil fine against the El Paso Energy Pipeline Group for safety violations.

Pipeline operators do most of their own inspections and develop their own safety standards, and they're not required to inspect regularly. And federal pipeline rules are so lax that, for instance, pipeline operators aren't even required to close their pipeline spigots when they find leaks. And they have an incentive not to, given that their income depends on keeping those spigots open and flowing. Many of the nation's pipelines aren't regulated at all by the feds--such as gathering lines that take crude oil to refineries, which Congress exempted from OPS oversight. Most of those lines didn't traverse heavily populated zones, so why bother protecting nature if nobody lives there?

"There are around 60,000 miles of unregulated pipelines and they leak like crazy," says Bob Rackleff, a commissioner of Leon County, Fla., who has spearheaded the National Pipeline Reform Coalition to push tougher regulation.

It generally takes a major explosion for the government to get involved. "There is almost no enforcement," says Lois Epstein, an engineer who followed pipeline safety for Environmental Defense in Washington, D.C., for 13 years. "Companies decide the appropriateness of their own lines with no federal standards."

Boiled in Oil

Even a horrific accident, though, doesn't seem to be enough to spur OPS--or Congress--to make meaningful change to protect the public from dangerous pipelines, which are increasingly turning up under heavily populated areas. Just a year before the New Mexico explosion, in June 1999, in Bellingham, Wash., Wade King and his buddy Stephen Tsiorvas, both 10, hiked down to a local creek where they found a lighter. As 10-year-old boys will do, they started to play with it.

Unbeknownst to the boys, a 33-year-old pipeline carrying jet fuel, gasoline, and other petroleum products lay directly underneath the boys' neighborhood. It had already ruptured, releasing an estimated 236,000 gallons of gasoline into the tributaries of Puget Sound, Bellingham Bay, and the boys' fishing creek.

A spark from the lighter turned the entire creek bed into a tunnel of fire, boiling the fish in its water. The boys jumped into the creek to escape, but the creek itself was on fire. By the time King's father ran down to rescue them, the boys had lost all of their clothes and most of their skin to the fire. Both died shortly after being airlifted to a burn center. Eighteen-year-old Liam Wood was also overcome with gas vapors while fishing. He fell into the creek and drowned.

Run by the Olympic Pipeline Company, Inc., the very same pipeline had leaked at least 47 other times since 1965. So far, the spill has been attributed to "mechanical failure" in the ruptured line, but the cause is still under investigation and points to other wrongdoing by the pipeline operator. Employees for the firm took the Fifth to avoid answering questions about just what went wrong.

The Bellingham accident did result in some action by the government. After two years of investigation, on September 13, 2001, the federal government indicted the Olympic Pipeline Company and three officials on charges of violating the Hazardous Liquid Pipeline Safety Act and the Federal Water Pollution Control Act. The case marks the first time the act resulted in criminal indictments. But tellingly, the impetus for the prosecution came not from OPS but from the Justice Department.

The accident and the ensuing investigation received widespread media attention, and brought calls from Congress for more regulation of the pipeline industry. The investigation also prompted unprecedented criticism of OPS. During the investigation of the pipeline explosion, National Transportation Safety Board (NTSB) Chairman James Hall said, "We believe [OPS'] lack of action continues to place the American people at risk."

Indeed, for years Congress and the NTSB had been asking OPS to require pipeline operators to radically improve inspections of equipment and training of employees who monitor the pipelines. But OPS has dragged its feet. OPS' record in adopting safety recommendations makes the Federal Aviation Administration (FAA) look like the gold standard--even after September 11.

In fact, OPS has the slowest and lowest record of any federal agency at complying with NTSB recommendations over the last 30 years. The agency complies with fewer than 70 percent of the NTSB safety recommendations, including many of those that would have prevented the accidents at Bellingham and New Mexico, according to Hall. (By comparison, the FAA complies with about 82 percent.)

For instance, NTSB wanted the pipeline office to require operators to perform their own inspections or tests in 1987. In 1992, Congress mandated that OPS require such inspections by 1995. But OPS only started requiring them for large liquid pipeline operators (those with 500 miles or more) last February--almost six years after the legislative deadline. (By law, government agencies must respond to but not necessarily implement NTSB suggestions.)

OPS hasn't even been able to keep up with the rules it's required by law to implement and enforce. As of May 2000, OPS had not implemented 22 statutory requirements and 39 of NTSB's safety recommendations--some going back to the 1980s, according to the General Accounting Office (GAO). A year later, GAO looked again and found OPS still hadn't met 16 of the legal requirements. Meanwhile, OPS had only complied with one of NTSB's outstanding recommendations and hasn't met six new ones issued since May 2000. OPS says it plans to comply with more NTSB regulations within another year, while it says others are so old they have become obsolete.

Stacey Gerard, OPS Associate Administrator, says that rather than respond to each of the proposals on a piecemeal basis, OPS is developing a long-term plan to deal with NTSB's concerns, a strategy she feels will ultimately be more effective.

But until some of those new rules come into effect, pipeline operators don't even have to inspect their lines regularly, let alone have the government do it or prescribe standards. And the new rules aren't exactly crushing. Once they take effect, pipeline operators will have a whopping seven years before they have to perform a baseline inspection (three and a half years for "high-risk" lines), and then they wouldn't be required to inspect more than once every five years unless the baseline inspection found a problem. And those rules only apply to large liquid operators. Inspection rules for small pipelines and gas operators are still being finalized.

The impact of this slow pace of oversight is clear: Despite all the publicity after the Bellingham accident and the calls for improved safety and better inspections, pipeline accidents have only gotten worse. According to OPS's annual figures (which are subject to change as more data come in), pipelines in 2000 suffered their worst year in terms of causing economic damage--$156,925,184 worth to be exact--nearly double the $86,856,82 OPS totaled for 1999. (Two major spills caused the blip.)

And those figures understate the problem. Companies aren't required to file reports on pipelines not subject to federal regulation. Given that the companies stand to incur fines for safety and environmental violations stemming from spills, they don't report them all. When NTSB investigates or someone sues, evidence often shows the oil companies underreport the problem, usually by understating the property damage. (The companies usually count mainly the damage to their own property.)

OPS also can't know of all the leaks and damage because operators don't have to report spills of less than 21,000 gallons (50 barrels). Bob Rackleff recalls that a now-defunct 85-mile pipeline running through the Florida Everglades between the 1960s and 1990s spilled 54 documented times, as reported to state agencies. But OPS files listed only a single leak.

Going With The Flow

Jim Pates has experienced OPS' indifference to pipeline safety and environmental concerns firsthand. Pates is the city attorney in Fredricksburg, Va., which he says has "the dubious distinction of having twice lost our entire water supply because of pipeline accidents." In 1980 and 1989, the Colonial Pipeline ruptured 30 or 40 miles up the Rappahannock River. Each time, hairline fractures grew over time until the pipe burst.

"After the second time, we decided to take a serious look at things and see what we could do," Pates recalls. "We saw OPS was not terribly concerned. They admitted to us after the 1989 break that environmental protection was not their thing."

Though Fredricksburg lost water for about a week each time, the pipeline still runs. OPS ordered Colonial to reduce pressure in the line, but still won't make the company fix it and Pates fears another crack could cause another water outage. "From all my years of being involved in government, I don't think I've ever seen an agency any more beholden to the industry it is supposed to be regulating," he says.

Part of the problem is that Congress simply has never given OPS the resources to do its job properly. OPS employs only 97 people, including 55 inspectors--about one per state and territory, or one nose per 36,000 miles of mostly hard-to-notice underground tubing--and even fewer experts to pursue enforcement when it finds problems. When asked by NTSB whether the agency had the capacity to enforce the rules it's writing, Clinton administration Research and Special Programs Administration Chief Kelley Coyner, who oversees OPS at DOT, replied "We do not. With 55 inspectors on hand, we are not currently in the position of reviewing these plans on a two-year basis to ensure compliance."

With so few resources, OPS has relied on the industry to perform its own inspections and develop its own safety plans. Rather than fining operators who run afoul of safety and environmental rules, OPS is now shifting its emphasis toward helping them find and fix problems.

There's plenty of evidence that this all-carrot-no-stick strategy will do little to protect the public.

Back in 1990, OPS assessed fines in almost half its enforcement actions. By 1998, that figure had plummeted to about 4 percent. Not surprisingly, over that same period, the number of deadly accidents has continued to rise. GAO reported last year that the number of "major pipeline accidents" (defined as those causing a casualty or at least $50,000 in property damage) rose an average of 4 percent a year between 1989 and 1998. OPS, however, attributes the increase to more lines, more flow, and more population near the lines.

The need for better pipeline oversight is growing--the pipeline infrastructure, to say the least, has reached senior-citizen status. About a quarter of it was laid 50 years ago or more. Sometimes the only way to measure flow is to count what goes in and what comes out, which doesn't account for volume or temperature changes or what might remain in the line. As a result, leaks on some of the older and smaller lines can continue indefinitely before anyone notices. Even detection systems of the 1980s are obsolete.

OPS official Byron Coy testified to NTSB that "fundamental systems that were installed in the '60s typically had useful lives of nearly 20 years, but the pace of technological advancement has caused some systems that were installed even in the '90s to be considered mature. Computer systems employed in many large pipeline control centers are very sophisticated, yet there are many small pipelines across the country that are virtually operated by hand."

Without the big stick of the federal government, pipeline owners have little incentive to upgrade their safety inspection systems. To fill the vacuum left by OPS, GAO recommended last year that OPS get states more involved in regulating pipelines to better leverage federal resources, but so far, OPS hasn't done much on that front. OPS already relies on 380 state inspectors who do most of the work that the companies don't do themselves, and states can use federal matching funds to regulate pipelines, but OPS has recently denied many states' requests to do so, and it hasn't been able to provide authorized matching funds.

The Piper Paying

That places the pipeline problem squarely in the hands of Congress, which has dabbled with it since the tragedies in New Mexico and Bellingham. But so far, Congress hasn't been able to pass even inadequate legislation. In 2000, the Senate passed pipeline safety legislation but since the House didn't move on it before Congress adjourned, the legislation died. Early in February of 2001, the Senate unanimously passed a similar measure but the House still hasn't scheduled hearings.

House Republicans possess plenty of incentives for not wanting to put damps on the pipeline industry. In the 2000 election cycle, the oil and gas industry gave $33,489,164 in campaign contributions, 78 percent to Republicans and making the industry the 9th-largest contributor by trade out of more than 80.

And who got more money than any other House candidate? Don Young (R-Alaska), chairman of the Transportation and Infrastructure Committee, who comes from a state with the largest number of pipelines. He got $133,600 for his last campaign from an industry that doesn't want the regulatory oversight the Democrats on his committee favored. Rep. Tom Petri (R-Wisc.) of the Highways and Transit panel, House Energy and Commerce Committee Chairman W.J. "Billy" Tauzin (R-La.) and House Subcommittee on Energy and Power Chairman Joe Barton (R-Texas) all have received hundreds of thousands of dollars in campaign contributions from the oil and gas industries.

The money has paid off. Just one example of the industry's clout: The Oil Pollution Act of 1990, passed in the wake of the Exxon Valdez oil spill to force oil tanker operators to pay for cleaning up the messes they make, exempts pipelines. Now, if an oil tanker explodes, the company has to clean up the spill. But a pipeline owner doesn't have to.

There is legislation afoot that would tighten federal oversight of the pipeline industry. In the transportation appropriations bill awaiting Bush's signature, OPS got funding for 26 more professionals it wanted, including inspectors and legal staff.

In addition, a Senate-passed bill, brainchild of Sen. John McCain (R-Ariz.), would require pipeline operators to develop safety plans, evaluate risks in heavily populated and environmentally sensitive areas, and educate the public about safety matters. The Senate Democrats' national energy strategy bill would also give DOT $3 million and five years to develop a research program to improve pipeline safety.

In the House, Rep. Jim Oberstar (D-Minn.) introduced a tougher measure that would allow states to regulate pipelines if DOT approves their plans, and give them about $20 million a year to enforce them. Meanwhile, Rep. Jennifer Dunn (R-Wash.) introduced a measure that would call for government--not industry--inspection and allow for a state role.

If these measures were backed with significant funding, they could go a long way to improving pipeline safety, but they still leave critical areas unaddressed, particularly the structural issues that make OPS so ineffective, such as the fact that the agency resides in a department mostly concerned with moving people rather than with enhancing safety and protecting the environment.

Though the federal government regulates pipelines because of environmental hazards, DOT remains in charge of oversight--not the Environmental Protection Agency (EPA) or even the Department of Energy (DOE), where expertise in preventing pollution and harvesting safe energy lie. That's because Congress established OPS before it created EPA or DOE and put it in DOT because product is transported across state lines.

The current legislative proposals also don't reconcile the conflict of who should pay for regulation. Right now, the oil and gas industries pay for their own enforcement. Though Congress must approve disbursements every year, OPS' budget comes mainly from two funds: the Pipeline Safety Fund, paid by interstate pipeline operators' user fees, and the Oil Spill Liability Trust Fund, which comes from spill damage recovery payments--another incentive for the industry to keep mum about spills.

Until these underlying issues are resolved, it's likely that the pipeline industry will continue to police itself, and innocent people like Amanda Smith will continue to die. It took the deaths of 3,000 people on September 11 to get the FAA to force decades-old safety regulations on the airline industry. We can only hope it won't take such mass casualties for the government to do the same for our pipelines. Too many people have died already.

Charles Pekow on Environment Reporting

Q: What drew you to environment reporting?

A: "Before I even read a lot of news articles, I was very interested in environmental issues. We gotta protect the world. When I was young, before I ever wrote a lead or news story, I would read a lot of environmental magazines and newsletters, like *Clear Creek*. I was interested in pollution, in the wasting of resources, air and water pollution, mercury in the fish – a big topic back around 1970. It was important. If we don't take care of the environment, what do we have left? We poison ourselves!

"In retrospect, many of the sources were biased. I was getting one side of the issue. At the time, though, I thought they were telling the truth. When you're younger, it's easier to get bogged down in perspective. Now, [looking back] I have to wonder if there was really as much poison in the food, if it really was as dangerous as they thought it would be. People were predicting mass starvation and population doomsdays. But what they were talking about and what I was worried about didn't materialize. I was a teenager when I was reading this – maybe 15-16-17 year old.

"With the combination of my training as a journalist and the wisdom of age, I can see things more clearly. Plus, more facts come out with time. But there's nothing more important than protecting the world, saving the world. We all need the environment. It should be in good shape, or we can't have good lives. The world will fall apart if we don't take care of the environment.

"The environment is the most important subject to me. It's closer to ultimate truth. [He cited some of philosopher

Martin Heidegger's views as well as Ludwig Feuerbach's] who said man arose from nature, the environment. It was a novel suggestion at the time. You gotta remember our roots. That makes it the most important beat."

Q: What makes environment reporting meaningful for you?

A: "A good story is a good story; but the environment was something of a tremendous concern for me. I wasn't equipped to be a scientist. I was equipped to be a reporter and an *investigative* reporter. So this is the one way I have of protecting the environment outside of doing the normal things of living an environmentally sound life.

"If there's any such thing as ultimate good or truth, it would have to do with the purity of resources and the purity of the environment. What right do we have as humans to despoil the Earth for other humans or other beings or for the Earth itself? There needs to be some environmental control. If that constitutes advocacy, so be it.

"Poisoning things in the wrong place can poison fish, people, can ruin the environment, can knock out invasive species. I always wondered, what right do we have, we're just humans, just one species, what right do we have to knock out other species for our own convenience. I feel quite strongly about that. Keeping resources pure strengthens diversity.

"I wonder now if we're getting too far away from ultimate truth with too much technology. Is life becoming too comfortable? Those are good questions that go beyond the scope I write about. They go into philosophy. The environment is everything. It's where we live. If we don't keep our environment in order, our lives eventually will become unpleasant.

"The environment is probably the most important because it is the one that transcends mankind, that transcends everything we do, where we came from. Without the environment, where would we be? It's something we have to watch, and clean up.

"Environment reporting is more likely to affect me personally. If I'm writing and exposé about some organization, the connection is a little less fluent. Environmentalism can mean two things: maintaining a world where people don't get sick, and it can also mean protecting the environment, because there's something more than mankind."

Q: What makes investigative reporting meaningful for you?

A: "I got into journalism and investigative journalism because I wanted to find truth. At some point I began to wonder if the journalism I was seeing was completely the truth itself! Or was it one more step in bringing the truth? With the communists, people thought they were creating the perfect society, and when they found out what it was really like, kaboom!

"I was in a school for disturbed children where the place was so restricted that we weren't allowed to read newspapers. So for a long time I didn't know what was going on in the world. I sort of became skeptical. I was also very critical of the press because sometimes when I'd see stories, they had nothing to do with reality. I experienced injustices...and truth in the world not being what [people say] it is.

"[Investigative reporting is] not a very lucrative way of making a living, but it's a way to be a good citizen. It's a way of using my ability to help improve the world. I get a little money, but more importantly I get the satisfaction of knowing I made some contribution, and I hope that it's for a better, more livable world for me, too. If I can maybe in some small way get people thinking about making pipeline safety better, it protects the environment for me, too."

Q: How do you view investigative reporting?

A: "A report covers. And investigative report uncovers. There wasn't much investigative reporting in Chicago's mainstream press when I was growing up. There is a fine line as to when good reporting becomes investigative reporting. It's a matter of art.

"That doesn't mean anything going beyond "he said, she said" reporting is investigative. Investigative reporting is taking something not handed to you. [Although that's also the definition of enterprise reporting, investigative reporting goes further, since enterprise reporting can also be in the "he said, she said" style, Pekow suggested, stressing again that investigative reporting uncovers, while an enterprise report can just cover.] Investigative reporting also must involve some wrongdoing, such as an abuse of public trust or a social or environmental problem.

"My biggest inspiration was Watergate. I also remember Jack Anderson's column in the *Chicago Daily News*. He was always coming up with big exposés. I also liked Ralph Nader and his "raiders," who wrote exposés about government programs, fraud and business. Those were major inspirations to me. Also, Mike Royko's columns of cronyism in the Chicago government. Upton Sinclair's *The Jungle* also had a big impact on me. It was a major, moving piece of investigative reporting, although it was fictional.

"Originally, investigative reporting relied on anonymous sources. At the beginning there was the image of the reporter as detective. That's what I thought investigative reporting was: a lot of shoe leather. Later on I thought it was going into government files. But initially it was just interviewing. Initially, when I read the reports or saw them on TV, I didn't think too much about the techniques. I though, you gotta get people to talk who don't want to talk. I've done that many times, and even given anonymity. But later on in my career I got to thinking that maybe journalists do that too much, and became more careful about that.

"My attraction to investigative reporting is philosophical. It's the idea of making the world what it says it is, of trying to get the reality behind the reality. So it's after the truth behind the truth, which people have been thinking about since ancient Greece and more recently since Hegel – you know, what the world *is* as opposed to what we simply see or what someone wants us to see. I wanted to see if I could find the truth behind the "truth," reality behind "reality."

"One doesn't need any particular aptitude to do investigative reporting. The most important thing is to be able to stick to a story. You have to be able to get doors slammed in your face and take it. You have to be able to call 12 people before finding one person who will talk to you. And sometimes you have to try different tactics on a person who doesn't want to talk to get the person to talk to you. Not many people can spot a needle in a haystack. But an investigative reporter can. It takes persistence rather than intelligence. And time. That's what investigative reporting has to do with aptitude. You don't have to have a specific degree or course of study. There are many people with degrees in journalism who are not successful at it [investigative reporting], and there are many people who are. I know people who don't have a college degree who do this type of reporting. On the other hand, a law degree could help. Or a degree in biology might make it easier to understand environmental stories better."

While Pekow said he has never plied a source with liquor to get the person to talk, he has traded information for that purpose. He also stays away from writing stories based on stolen memos. "[Yet] people have given me things that they didn't mean to and said 'whoops' and I say, 'Too late!' You don't go through someone's legal files when you're in their office and left alone, but if it's left on the desk and you can look at it, yeah. But I don't remember getting something useful that way."

Pekow added that from investigative reporting in general he gets a sense of spiritual wealth. "It inspires me, but it's much harder to sell."

Q: What criteria do you use for moving the public with your investigative reports?

A: "It's finding something that's not obvious and going out and digging, through files, finding recalcitrant witnesses, not taking the first answer. Asking the tougher questions and delivering an artful telling of the story."

Q: How and where do you draw the line between advocacy and environment reporting?

A: "I remember reading one article that said the way to stop air pollution is to stop driving cars. The way to do that is to pour sugar syrup in the gas tanks. That was *Clear Creek*. So they were not just reporting, they were encouraging action.

"Advocacy journalism advocates a point of view or an action. I can advocate a legislative or regulatory action, or that people carpool or boycott a product. As long as it's written based on facts, it's advocacy journalism. Everything from *The New Republic* to *The Nation* to *The Weekly Standard* is advocacy journalism. For that matter, Sierra magazine also. They're asking people to take positions and write members of Congress – it's definitely writing with a point of view."

Q: How and where do you draw the line between advocacy and investigative reporting?

A: "Advocacy journalism belongs anywhere, as long as it's labeled as such. But it doesn't belong in a nightly newscast. And if it appears in a magazine or newspaper, it belongs in the opinion section or the editorial or op-ed pages. It should be clearly labeled as opinion.

"Advocacy is that which includes a call to action or point of view, for instance, what a specific policy should be about something. Slant is in the eyes of the beholder. Slant is when you're not being fair to one side. In advocacy, you're making it clear you're promoting something. With slant, you're just leaving out part of the story but making it appear that you are not. In that case, you're being slanted and appearing to be objective. Modern journalists have been trained not to be advocates.

"It is a weakness of mine, and one of the greatest weaknesses of man [sic], that we tend to want to read what fits our own prejudices. I'm concerned about this, personally. The problem with advocacy journalism is they're preaching to the choir, largely, Pekow said, implying non-advocacy journalism reaches a broader audience."

Q: What are some realities of the environment beat?

A: "You have to humanize the stories, make them readable by talking about specific cases. That makes them interesting to read," Pekow said. [In his experience, reports about the environment sell just as easily as any other story, he said.]

Q: What does the future look like for environment reporting and for investigative reporting?

A: "The environment beat is probably going to continue to be fractured in the sense that a lot of it is going to be done by specialized publications – a newsletter covering sludge hauling or propane, for instance. So the only people who will be reading it are people with a specialized interest in that one particular area. The environment is not a big story for he mainstream press, unless it hinges on another beat: war, politics, etc. It's still somewhat of a lower level in the mainstream media. Unfortunately, I don't know that that's going to change. In the mainstream press now, I see a smaller quantity of reporting on the environment."

Pekow also cited media ownership concentration as presenting a conflict of interest when it comes to reporting about the environment. Several large owners of media conglomerates are "in the energy business or the land use business. Does this mean they're going to pull their environment stories? What it may mean is shying away from environment stories. Is ABC going to want to cover a land use story [fairly] when Disney (which owns ABC) wants to knock down trees and open a theme park? Synergy will definitely become a problem. General Electric owns NBC as well as energy and engineering businesses that affect the environment.

"How is Westinghouse going to cover an energy crisis? It's something [the media and the public] are going to have to keep an eye on. Is the environment beat really overblown? Should it be discussed? How clean is clean should be debated. And, how do you communicate risk? What environment stories do you cover?

"I am willing to accept criticism about why I am writing stories about recreational trails. I consider that environmental because it has to do with transportation. Why am I concentrating on stories for bicyclists as opposed to something else? Because bicycling is an interest of mine, and because [some of] my clients are bicycling magazines. What's saleable has a lot to do with the future of investigative reporting about the environment."

Charles Pekow's Reporting Process

In the January/February issue of *The Washington Monthly*, Pekow delivered a gripping story of a family cremated while camping along the banks of the Pecos River in New Mexico when a corroded oil pipeline "ruptured and sent a flaming fireball across the campground. Burning fuel rained down upon the campers. The fire was so hot it melted sand into glass and turned parts of the concrete bridge into powder. Tents and sleeping bags turned to soup. The flames leapt 500 feet in the air, were visible from 30 miles away, and left a crater 86 feet long, 46 feet wide, and 20 feet deep.

"Amanda Smith survived the fire only long enough to describe the horror of the scene to rescue workers. The explosion wiped out most of the Smith family, killing Amanda, her parents, her husband, her two kids, her brother and sister-in law, their 22-month old daughter and twin 6-month-old babies," Pekow wrote.

Pekow described how he collected that report. "In environment reporting, you're obviously dealing with something scientific, something harder than if you were doing something strictly human [as in 'human behavior']. So you may have to find more physical evidence or documents than if you were reporting on some other form of corruption or dishonesty, where you'd have to interview a lot of witnesses. The scientific expertise makes it more difficult to do if, like me, you don't have that. I have had to familiarize myself more with scientific issues. The big help to me was knowing something about the environment and the drawback was what I didn't know.

"What I learned through beat reporting was somewhat inadequate [when it came to reporting about the environment]. I had to do a lot of background research. On the pipeline story, I knew, for instance, about the pipeline safety program. I learned through beat reporting that the government's efforts to regulate pipelines were inadequate, that the government did not have the resources or an effective way of protecting the public and the environment from pipeline tragedies.

"By beat reporting, I mean just covering a topic. I write a monthly column on propane for a propane magazine. In the course of writing that column, I came across a problem I thought deserved some investigation. So I went to it!"

The story became Pekow's first in-depth report involving the environment. "First I went through government documents. Some of the things were in the Federal Register. And I had to dig through the Office of Pipeline

Safety's information. Some of it's on the Web, but you have to look for it. You have to look very carefully and deeply at it and through a bunch of other government files.

"I then interviewed some people – experts in the oil business, and one person in the environmental movement in Alaska who I was able to identify as an expert on pipelines. So I had to track her down. I also interviewed members of Congress. I picked up newspaper clips from way back when, looked into the National Transportation Safety Board and pipeline safety, rummaged through files – it's easier now than it used to be because so much of it is on the Web – and put together a story that put two and two together."

And, Pekow adds, "I had to get some scientific facts in addition to quotes. I had to document what the dangers were about corrosion in pipelines and so forth, and the actual dangers of them exploding."

Chapter 5: Paul Rogers

Paul Rogers, born in England to British parents, moved with his family to New York City at age two, and later to Ohio where he grew up. He obtained two bachelor's degrees from Indiana University – one in journalism and the other in political science. Rogers is now the natural resources and environment reporter for California's *San Jose Mercury News*, a position he has held since 1995.

Rogers is responsible for raising regional and state environmental concerns, and occasionally national ones. Before gaining that position, he covered coastal issues from the paper's Santa Cruz bureau. Rogers holds a Pulitzer Prize for team coverage of the Loma Prieta earthquake in 1990 for general news reporting (an explanatory report).

Rogers said he has no interest in becoming an editor. He is also not excited about doing investigative reporting, although he is crazy about the results it yields. Environment reporting is "fun" and investigative reporting "is a drag," he said, then qualified his statement, "[Investigative reporting] is exhausting. When I was in the middle of the investigative report about cow grazing, I wanted to quit. I was miserable. And I realized, this just isn't me," he said.

He continued, "There comes that awful moment when you have to sit down and write. And the more you've collected, the higher the drop is going to be. And you're miserable." For the change it prompts, Rogers endures investigative reporting's tedium, but not gladly. His investigative reports about the environment typically focus on politics and injustice.

For three years, between 1999 and 2002, Rogers led classes at the University of California-Berkeley as its Hewett Teaching Fellow in environmental journalism. In 2000, he became a lecturer in the science communication program at the University of California-Santa Cruz, a part-time position he continues to hold. In 2003 he received the Sierra Club's David R. Brower Award for environmental journalism.

Cash cows: Tax dollars still support a Wild West holdover that enriches big ranchers and degrades the land. Defenders say cattle jobs are rural lifeblood.

Published November 7, 1999, *San Jose Mercury News*
By Paul Rogers and Jennifer Lafleur

By many measures, the American West at the end of the 20th century is more environmentally healthy than a generation ago. The air is cleaner. New national parks flourish. The bald eagle and California gray whale are back from the brink of extinction.

But on millions of acres of public land from the Mississippi River to the Pacific Ocean, one practice still perpetuates environmental damage that began in the Wild West 150 years ago -- cattle grazing. Propped up by more than $100 million last year in taxpayer subsidies, ranchers leasing national forests and range managed by the U.S. Bureau of Land Management (BLM) have allowed millions of cattle to grind down native grasses and trample streams, the lifeblood of the West. Some wildlife have even been driven to the endangered list.

Yet because hoofprints in streams aren't as dramatic as oil spills, most Americans aren't aware of the damage.

Even fewer realize they pay corporations and millionaires to create it, all within the law.

A nine-month Mercury News investigation has uncovered the extent to which wealthy hobby ranchers, agribusiness giants and corporations -- far more than hardscrabble families portrayed in John Wayne movies -- benefit from federal grazing subsidies.

Through Freedom of Information Act requests, the newspaper obtained billing records of the more than 26,300 livestock operators who lease land from the U.S. Forest Service and BLM. An analysis found such Rolex ranchers as hotel mogul Barron Hilton, beer giant Anheuser-Busch Inc. and Mary Hewlett Jaffe, daughter of Silicon Valley billionaire William Hewlett, enjoying below-market grazing fees.

"The only justification for this subsidy is folks wrapping themselves in the cloak of the family farm," said Thomas Power, economics chair at the University of Montana. "Once it becomes clear these subsidies are primarily going to large corporations, the cover is blown. It's politically obnoxious. We're rewarding the rich at the cost of environmental damage."

In 17 Western states, livestock grazing is allowed on 254 million acres of national forests and BLM land -- an area equal to California, Oregon, Washington and Idaho. On that vast expanse, about 26,300 ranchers graze 3.2 million cattle.

When it comes to grazing at the federal trough, no one sits taller in the saddle than corporate cowboys. Their cattle -- and to a lesser extent, sheep and horses -- dominate the vast public lands of the West.

The top 10 percent of grazing-permit holders control a striking 65 percent of all livestock on BLM property, federal records show.

Similarly, on national forests, the top 10 percent control 49 percent of the livestock. Many of the largest are grazing associations in the Dakotas, Wyoming and Colorado, made up of family ranchers. Others are corporate entities such as Idaho billionaire J.R. Simplot, media mogul Ted Turner or Hunt Oil Co. of Dallas.

Why do wealthy interests run cattle at all?

Some executives use ranches to meet with major clients. Companies also buy ranches for water or mineral rights, and raise beef to offset expenses. For some rich Americans, ranches are status symbols and cows are part of the image. Others are simply in the beef business.

Defenders say it doesn't matter if corporations dominate public land.

"Yes, they make a lot of money selling computers or beer or potatoes, but they're still running the same cattle, and they still employ the same number of cowboys and ranch hands," said Jason Campbell, federal-lands director for the National Cattlemen's Beef Association, in Washington, D.C. "They are supporting mainstream businesses in places like rural Wyoming and Colorado and Montana."

Reform, they say, would burden an industry already challenged by low beef prices, environmental lawsuits and foreign competition. And, some contend, it would threaten rural banks that carry loans pegged to ranchers' grazing permits.

The red ink runs deep, however, to keep cows on the public's range.

Last year, the forest service and the BLM lost more than twice as much money on grazing programs as they spent to restore endangered species. Together, the agencies lost $94 million on grazing, spending $116 million and taking in only $22 million in fees. Under another program, known as Wildlife Services, the U.S. Department of Agriculture spent $14.6 million more in state and federal funds to kill coyotes, mountain lions and other predators for Western ranchers.

It's not as if Americans need the meat. Only 3.8 percent of the nation's beef cattle graze on federal lands, according to the Department of Agriculture. The rest live on private ranches, mostly in Texas, Oklahoma, Nebraska and the Dakotas.

Reform has been slow. Few politicians want to take on the West's ultimate hero, the cowboy. Ironically, the small ranchers he represents have a tiny piece of the public range. The Mercury News found that the bottom 50 percent of grazing permit holders on national forests control 3 percent of the livestock; on BLM land, the bottom 50 percent control just 7 percent.

Six years ago, the Clinton administration launched attempts at change. Some members of Congress also have made futile efforts. While some new environmental protections have been put in place, much has stayed the same. Consider: the fee ranchers pay to graze livestock on federal land is lower than at any time since 1975, set at $1.35 per cow per month, far below the current market average of $11.10 on private land in the 11 westernmost states.

Under a "use it or lose it" permit system, a small group of ranchers enjoys a monopoly on federal lands, as they have for decades. Even when hunters, fishermen or environmentalists have offered to outbid cattle operators and remove cows, a 1934 law prohibits it.

Despite the Clinton administration's promises to better balance wildlife, recreation and grazing, the number of cattle and sheep roaming federal lands has changed little since the Reagan presidency.

"One very small, politically powerful industry is destroying our land," said John Horning, director of watershed protection for Forest Guardians, a Santa Fe environmental group. "But the salt in the wound is that we're paying them to do it."

French-fry billionaire Idaho tycoon has access to millions of U.S. acres.

The king of public-lands grazing is an Idaho billionaire who provides half the French fries for McDonald's restaurants nationwide.

John "J.R." Simplot, 90, is listed on the Forbes 400 list with a net worth of $3.6 billion. Ranked by cattle numbers, he reigns as the largest operator on BLM lands in the United States. Simplot controls permits to graze on at least 1.9 million acres of BLM land -- an area six times the size of Los Angeles -- mostly in Idaho, Nevada and Oregon.

Simplot may be the only American to have made a billion dollars in both 19th-century and 21st-century businesses.

His holdings include food processing, fertilizer and other divisions making up the J.R. Simplot Co., a private company based in Boise. Simplot also was an early investor in Micron Technology -- his family and companies control 12 percent of its stock, worth $2.3 billion.

Tom Basabe, 47, president of Simplot Livestock Co., based in Grand View, Idaho, said the company takes good care of public lands. "We make our living off the land," he said. "We would be very foolish and shortsighted if we abused it."

Government already imposes too much paperwork and too many regulations, Basabe said, and with Interior Secretary Bruce Babbitt and the Clinton administration, the bureaucracy has worsened. On this day there are 100,000 cows just outside Basabe's office on Simplot's Grand View feedlot, the largest west of the Mississippi. Visitors can smell it three miles away.

In long rows of dirt pens and feeding troughs, the cows await shipment to slaughterhouses in Utah and Idaho. Some meat will be sent to Japan.

Basabe says big operators should continue to pay the same grazing fee as small family ranchers.

"Are you going to charge two prices for a loaf of bread?" he said. "It's unfair."

Chuck Jones, Simplot's public lands manager, said the company's cattle also reduce invasive plants and fire danger. And Simplot cowboys patrol vast open areas.

"We'll never apologize for being the biggest," he said. "We admire Mr. Simplot for being an American success story."

Trail to cattle country:
Why ranchers pay $1.35 a month per cow

There are few places cattle have not penetrated.

Introduced from Europe 500 years ago, cows today graze Southwest deserts. They wander red rock canyons of Utah and evergreen forests from the Sierra Nevada to the Rocky Mountains.

The basic grazing arrangement seems simple enough. Ranchers obtain 10-year permits for large blocks of public land known as allotments. The only requirement is a private ranch and livestock such as cattle, sheep or horses. Permits often stay in families for generations.

But the formula that establishes grazing fees is a complicated one. Factoring in beef prices and other costs, Uncle Sam charges ranchers $1.35 a month for each cow and calf.

That's 87 percent lower than the average rate of $11.10 that ranchers charge each other to graze private property in the 11 westernmost states. Some places in Santa Clara County charge $15.

Critics, who like to point out it costs more to feed a hamster than the federal government charges to feed a 1,000-pound cow, say the fee is a scandal. Some suggest it be quadrupled to ensure the government breaks even.

"Clearly, taxpayers are not getting their money's worth," said John Berthoud, president of the National Taxpayers Union.

Defenders counter that the fee simply reflects drops in beef prices and that few cowboys get rich.

"We are almost in a life-and-death situation here," said Caren Cowen, executive secretary of the New Mexico Cattle Growers' Association. "It's getting tough to be a cowboy."

Many ranchers say it's wrong to compare fees on private land -- where fencing, water tanks, veterinary help and other amenities are provided -- with public land, where they are not.

"You dump your cattle off and a guy takes care of 'em for you. He doctors them, feeds them and rides them. You just pick them up in the fall," said Dave Nelson, former manager at the Mountain Springs Ranch near Sun Valley, Idaho. The 14,000-acre spread was owned for 25 years by computer pioneers David Packard and Bill Hewlett until their families sold recently to Hewlett's daughter, Mary Hewlett Jaffe.

Ranchers don't have much choice, Nelson noted; in many Western states, more than half the land is owned by the U.S. government. Also, because it's public land, hikers, hunters and anyone else can wander among federal pastures, occasionally causing mischief.

"If I could rent pasture at $12 to $14 a month on private lands, I'd do it," he said. "It's like having a motel room with the door not locked. People get out there and cause problems, they cut your fences."

But nearly every state in the West charges more to graze state-owned lands than the federal government. Oklahoma, for example, charges up to $10.80. Montana charges $4.40 a month for lands, most adjacent to federal property.

Even some cowboys call the fee too low. Most live in the Great Plains, where nearly all land is private.

"We consider it to be an unfair subsidy," said Scott Dewald, executive vice president of the Oklahoma Cattlemen's Association. "The fee should be based on public auctions, high bidder takes it."

The $1.35 isn't the only subsidy.

Half that fee is given back to the ranchers as "range betterment funds" to pay for fences, water tanks and other equipment. When Anheuser-Busch cows are walking along a barbed wire fence in the Sierra wilderness, for example, the likelihood is tax money paid for it.

Some people argue that in some cases, the choice is cows or condos.

The ranching industry, top government officials and even the Nature Conservancy warn that if fees are raised to make corporations pay more, small operators will go bankrupt and real estate speculators will carve up the empty private ranches.

"What are the alternatives?" said Mike Dombeck, chief of the U.S. Forest Service. "Are we better off with subdivisions? I think its in everybody's best interests to try to maintain these large unfragmented tracts of land."

While cowboys dominate the mythology of the West, public-lands ranchers are a rounding error in the mathematics of the cattle business. They make up just 2 percent of America's 1.1 million cattle operators, according to the U.S. Department of Agriculture.

Yet the mythology continues.

Wyoming has a cowboy on its license plates. Yet Florida has more beef cattle. Nevada holds a cowboy poetry festival, but has half as many beef cattle as Louisiana.

And so on across the West.

The reason? Water.

It is easier to raise livestock in the East, where water and grass are plentiful, than in the arid West. A cow in the East can live on 2 acres. In the West, each needs up to 100 acres to find enough to eat.

Cattle-industry leaders say that even though public-lands ranchers provide a tiny amount of America's food, they are crucial to rural areas.

"Sure, it becomes a lesser issue when you frame it in the national context," said Campbell, of the National Cattlemen's Beef Association. "But if you go into some of these small counties, you quickly realize that if a lot of those ranchers didn't have access to federal lands, they wouldn't be there."

Their importance may be more legend than fact, however.

Power, the Montana economist, concluded in a 1996 book that only 17,989 jobs in 11 Western states, or 0.06 percent of total employment, were due to federal grazing.

"We have cowboy socialism," said Power.

"It's a romantically based, phony attempt to protect something from the past that no longer exists."

Hilton's second empire:
Hotel chairman uses vast tracts for herds

Among the most well-heeled cowboys is Barron Hilton.

Hilton, 70, is chairman of the Hilton Hotel empire and has a net worth estimated at $500 million. An avid fly fisherman and hunter, he owns one of the trophy properties of the West, the Flying M Ranch, about 50 miles east of Carson City, Nev.

The ranch sprawls over 450,000 acres, made up of private property and huge stretches of BLM land and Humboldt-Toiyabe National Forest. Hilton holds permits to graze 3,200 cattle all the way across the California line, nearly to Mono Lake.

He bought the ranch in 1969 for fishing, hunting, tennis and skeet shooting, said Steve Hilton, Barron Hilton's son, and more recently has used it for soaring and gliding contests. Running cattle helps offset operating costs, said Steve Hilton.

One family adviser acknowledged the $1.35 fee is a pretty good deal.

"Yeah, I think it is too low," said Don Hubbs, chairman of the Hilton Foundation in Los Angeles, and a rancher himself. "I think the fees could be raised. But a lot of this government land isn't as good as the private land."

Hubbs said public-lands cowboys have an undeserved reputation.

"The public has been made to believe that ranchers are anti-environmental," he said. "In a few instances, maybe that's true, but in the vast majority it's not. The true ranchers know that once you desecrate the land, you don't have the production from your cattle."

Where damage is done:
Stewardship hasn't ended destruction

Some environmental groups have argued since the 1980s that all cattle should be removed from public lands because of overgrazing.

Cowboys counter that most harm was inflicted more than 50 years ago. Back then, agricultural textbooks called streams "sacrifice zones," and many ranchers turned wilderness areas into de facto feedlots.

Today, ranchers say, they are doing a better job of stewardship.

Both sides are right.

A sweeping study by the Department of the Interior

in 1994 concluded that although overgrazing of "uplands" -- pastures, deserts and meadows -- has fallen in recent decades, stream areas continue to suffer severe damage from grazing.

"Cattle have been the most widespread, destructive influence on the streams and rivers of the West," said Robert Ohmart, professor of biology at Arizona State University. "They have just beat places to hell."

Small family operators appear to be taking better care of the public's land than the largest ranchers. *The Mercury News* found that 46 percent of the allotments rented by the 20 largest BLM permit holders are classified as "improve" -- unsatisfactory conditions in sensitive areas. In contrast, the BLM assigned that grade to only 27 percent of all its allotments.

Not all grazing is harmful. In select areas, it can limit the spread of non-native plants, helping wildlife. But a 1994 report by forest service biologists found grazing to be the main reason species are put on the endangered list in the Southwest.

Again, it all comes down to water.

Much of the region receives less than 10 inches of rain a year. Streams make up 1 percent of the landscape in the 11 Western states, yet 70 to 80 percent of plants and animals depend on them for survival.

Unlike elk or bison, which eat and roam, cows love to congregate in streams, especially in summer. If they are not herded or fenced out, they can wipe out fish and birds.

The news isn't all bad. Tougher rules from the Clinton administration have resulted in gradual improvements. In 1995, 27 percent of BLM streams were classified in "proper functioning condition," by the agency's scientists.

By 1998, that total had risen to 36 percent.

Wayne Elmore, an Oregon ecologist who heads the federal government's streams team, has conducted studies showing that streams and wildlife can recover, but usually if cows are removed for five years or more. After plants grow back, cows must graze only part of the year, he argues.

"We need better management," said Elmore, "not fewer numbers."

Compounding the problem is a lack of information.

Unlike the BLM, the forest service has no national database listing stream conditions, cattle locations or the names of people with grazing permits. That has made it nearly impossible to track trends. *The Mercury News* combined computer data from the agency's six Western regions to rank permit holders.

"We recognize that there are deficiencies in the databases. We are working on that," said the forest service's Dombeck, who supports higher fees and tougher environmental rules. "It's a few years away."

War in the courts:
Suits regain territory for endangered species

If anyone is going to kick cows off public lands, it may be environmentalist John Horning and his lawyers.

On a warm fall day, Horning is zooming through western New Mexico. Cibola and Catron counties are the kind of places where locals stock ammo for Y2K and pick-up trucks sport bumper stickers like "Sierra Club: Hike to Hell." Billboards advertise "custom slaughter and taxidermy."

There is boundless space, with hawks flying over tin-roofed houses and hazy hills 50 miles in the distance distorted by rising heat.

Near the Arizona border, Horning wanders the San Francisco River, a scenic watercourse through the Gila National Forest.

"All of this area we're walking on should be covered with cottonwood trees," he said. "But it's not. Because of overgrazing, it will take 60 to 80 years for this to come back."

Horning's employer, the Forest Guardians, has made its mark here.

In a landmark decision last year, the government agreed to remove cattle from stream areas on 57 grazing allotments on the Gila and Apache-Sitgreaves national forests in response to a lawsuit from Forest Guardians and other environmentalists. The suit alleged that grazing drove two fish -- the loach minnow and the spikedace -- nearly to extinction and harmed an endangered songbird, the willow flycatcher.

Taking on an icon:
Politicians, confusion stall significant change

Change has come slowly. Why?

Grazing damage is out of sight and mind for most Americans. The economics are confusing. Cowboys are popular. Also, banks use grazing permits as collateral; if the system changes dramatically, thousands of ranches plunge in value. Many bankers quietly lobby against change.

"If in the middle of a loan we have an increase in grazing fees, then that dips into a rancher's ability to repay his loan," said John Anderson, vice president of the New Mexico Bankers Association.

Backed by a sympathetic constituency on an issue

most Americans don't comprehend, a handful of Western senators, led by Sen. Pete Domenici, R-N.M., have been able to block efforts to significantly raise grazing fees.

Domenici declined repeated requests for an interview.

However, he has cited the family rancher in defending the status quo, even though 98 percent of U.S. ranchers don't have public-lands permits. In 1996, he told the Albuquerque Journal: "It's about whether you want to have any more cowboys around the West or if you want them all to come from Hollywood."

A Bay Area Democrat, U.S. Rep. George Miller of Martinez, has led efforts for nearly a decade to raise fees and tighten environmental rules. Little has changed.

"You run into a bipartisan bloc of Western senators that decide they are going to fight for this as hard as they fight for anything," Miller said. "They are fighting for an old, irresponsible use of land."

When President Clinton began his first term in 1993, he unveiled plans to reform mining, logging and grazing practices on public lands. He said the changes would save $1 billion over five years. Yet: Clinton quickly dropped a plan to raise grazing fees in March 1993 when five Democratic lawmakers joined leading Western Republicans in protest. Clinton decided that because of the rift, the five might not vote for his economic plan.

Critics called it a cave-in.

"Ultimately, it was the president's call. The economic plan was the last thing we wanted to go down in flames," said former White House Chief of Staff Leon Panetta, who added he still feels that "taxpayers are not getting a fair return here."

The same year, Interior chief Babbitt tried to raise

fees to $4.28per cow a month, but Domenici blocked it with a filibuster. After the GOP won Congress in 1994, the administration gave up on fees.

When BLM and national forests are combined, the overall livestock total has fallen just 1.5 percent since 1988. Critics say rural federal officials rarely remove cattle to restore the environment because of local backlash.

"The Clinton administration promised great strides; what we ended up seeing was a lot of baby steps," said Cathy Carlson, a grazing activist with the National Wildlife Federation in Boulder, Colo.

Today, Babbitt defends the administration's efforts.

He points out that he has put rules in place under an initiative called "Rangeland Reform '94" to keep cattle out of streams and to increase public participation in regional range-advisory councils.

"We took our best shot," Babbitt said. "We have made some important progress. But there aren't any magical breakthroughs. Every range reform of the past 100 years has played out amid intense controversy, but we are gradually making our way."

Some officials are more blunt.

"It's frustrating," said Jim Baca, a former BLM director whom Babbitt forced out in 1994 after run-ins with Western leaders. "It seems like nothing has changed."

Baca, now mayor of Albuquerque, said high-tech jobs, such as the new Intel Corp. plant near Albuquerque, provide far more economic power to the West than ranching. But old habits die hard, he said.

"When I was BLM director, if I wanted to move 10 cows in Wyoming, senators would get involved," he said. "I still marvel at it."

Even Babbitt's Rangeland Reform is in jeopardy. A

lawsuit filed by the Public Lands Council, a ranching group, would overturn it. Last month, the U.S. Supreme Court agreed to hear the key case.

So how does Uncle Sam bring back the environment without bankrupting family ranchers?

One idea is to buy out public-lands cowboys. The Grand Canyon Trust, an Arizona environmental group, recently removed cattle from 132 miles of streams in Utah's new Escalante-Grand Staircase National Monument by paying five families to drop grazing permits.

Terms were not released, but the trust said the buyout cost between $215,000 and $540,000.

Meanwhile, action by the GOP-led Congress to make corporate ranchers pay more is considered a political impossibility. Clinton could raise fees by executive order, but administration sources say he has decided the fight isn't worth it.

And the issue hasn't surfaced in the 2000 presidential campaign.

"Most people just don't care," said a Clinton administration source who requested anonymity. "It was so long and tiring and taxing before, I just don't know if we're going to see any more reform efforts. There's no upside to it. We used a ton of political chits and got nothing."

Paul Rogers on Environmental Reporting

Q: What drew you to environment reporting?

A: I wanted a job where I would get paid to go to Yosemite National Park. There are a lot of other jobs I could have done here in the last few years but that I've turned down. I could have been our main Washington, D.C. correspondent. I could have been in our Vietnam bureau. I could have been in our Sacramento bureau covering state politics. I could have covered the governor's race. The presidential race. You name it. I gave up a job where I was covering Monterey Bay and Santa Crus and Pebble Beach and Big Sur and Carmel [as municipal beats]. People go on vacation there all the time! There are a lot of other folks I know who would like to do that.

"But one of the things about the environment beat that's so wonderful is it's challenging, it's intriguing, everyday is something completely different. You get paid to go to all these interesting places, unlike the White House, where you're basically a stenographer following people around and regurgitating into a computer. [On this beat,] the stories you write make a big difference, because there aren't that many people on this beat. So, if I suddenly decide it's very important to start doing stories about invasive species in San Francisco Bay and all over the place, that becomes an issue and legislators will introduce bills. Or, if a decision is made that we're going to do stories about how the San Joaquin River and Sacramento River are polluted from farm run-off, advocacy campaigns are going to spring up around those from environmental groups, and lawsuits will be introduced, and things like that will happen. I could do this for another 30 years and it wouldn't bother me. There's such a learning curve on this beat.

"That's one of the satisfactions that comes from doing this beat. The longer you do it the more you can understand and the deeper you can bore into it, the better the stories you can write, and the more impact you have. [I] love the give and take, the ebb and flow of the democratic process. And the democratic conversations. It's nice to be an integral part of that, by putting issues on the table to discuss and fight about."

Q: What drew you to environment reporting?

A: Environment reporting has made Rogers less of an environmentalist than before he started on the beat, he said.

"I used to assume that environmentalists always have the best interest of society at heart, and they never lie to you, and they were always good guys. And I found that environmentalists mislead, obfuscate, and lie as much as government or industry does. I have learned to look at these issues with much more of a healthy skepticism,

"This beat also made me realize I don't want to be an editor. I like doing this too much. It's fun. I've climbed down the side of an oil tanker outside the Golden Gate Bridge on a rope ladder at midnight with 10-foot waves crashing against my feet. I've seen grizzly bears in the wild in Alaska. I've hiked through the Amazon rainforest with some of the best tropical biologists in the world. I've gone out on cattle pastures with Mormon ranchers in Utah. I've talked with tree sitters and squid fishermen and strawberry farmers and all sorts of interesting characters. It's a really fun way to make a living!

"The environment beat is among the most fascinating. I'm a political junkie. I love politics. I have a degree in political science. And I would rather do this than cover the White House."

Q: What personal lessons have you gleaned from investigative reporting about the environment?

A: "Investigative reporting is different than I imagined. It's harder. Many reporters begin with a theory and end up with nothing. There's very, very little about the process of investigative reporting that's rewarding. It's one of the most stressful, miserable things that I've ever seen. You run down hundreds of blind alleys, you spend many days in the newsroom until midnight going through thousands of statistics. You're threatened with lawsuits all the time. It's a very, very unpleasant craft, I think.

"Sometimes you just throw your hands up. But usually you come out of the nadir and write something you're proud of, that either does a great job or explaining a complex facet of society to the public – which is your job in a democracy – or you expose some level of corruption, or some level of unfairness in society that the public sees and grows from. In the end there's a change based on your reporting. That's what the benefit is. It's the kind of thing that a lot of people aren't cut out for. It's the highest and best use of your journalism career, but it's really difficult. I didn't realize that when I got into it.

"I was miserable! I had to move my desk to another part of the room to get away from distractions. And I realized that this just isn't me. I didn't realize that until I was deep into a project.

Q: What makes the environment beat meaningful for you?

A: "There are so few people who cover this beat all the time. The level of expertise is so high that you write with authority. You're much more likely, when you write a story on this beat, to influence policy, to write something that will get noticed, that will go out on the wires and have millions of people read it, that will shine a light on some problem, or expose some taxpayer rip-off. Environmental stories seem more likely than others to have people act on them. To get results. If you just write the truth, there are plenty of activist groups that will embrace whatever cause is in the story. You don't have to do their work for them. Just shine the light."

Q: What makes investigative reporting meaningful for you?

A: "It's something that feels meaningful; it's something where if you're an idealist you feel like you're not just making money for money's sake. You're not just screwing the tops on toothpaste tubes as they go by you on an assembly line all day. It's not a frivolous pursuit. It's not something where you feel you've wasted your time. Good investigative reporting creates bodies of new, original research. It turns out things that you can't find in any other report that you can't get anywhere else.

"I get something out of investigative reporting that environment reporting alone does not provide. It get the same satisfaction, but more so. If you like to get paid to learn, you're getting paid to become a real expert. If you like to get paid to influence society, hard-hitting investigative pieces really influence society. if you like to get to do a job where you can see the results of your work, it's likely that at the end of an investigation someone is going to introduce legislation or do something to make the change. Investigative reporting is regular daily journalism with extra stresses and extra rewards."

Q: How do you view investigative reporting?

A: Rogers came to investigative reporting with few ideas about what it should accomplish. Growing up, he did not read investigative reports or watch them on television, nor were they discussed over dinner or at any other time in the household. Rattling off historical case examples and statistics, he demonstrated his passion for facts as a means for criticizing public policy and illustrating injustice.

"Getting the facts is how a responsible journalist locates and reveals truth. Investigative reporting must cause outrage in its audience to be worth anything. The very best investigative stories combine the outrage of some kind of environmental degradation with the outrage of taxpayer rip-off. There's nothing more fun to me than to watch lefties and righties get together on an issue, to see people from two ultra conservative political policy think tanks - the Cato Institute and the American Enterprise Institute – agreeing with the Sierra Club and the Natural Resources Defense Council. And it happens. It happens more than you think. Those are the best environment stories.

"In the beginning of each year I make a list of stories I want to do so I don't end up doing 100 15-inch stories and never do anything with depth that year."

Q: What criteria do you use for moving the public with your investigative reports?

A: The outrage factor is one of five criteria Rogers uses to determine whether an investigative report about the environment will move the public. Together, they are: (1) How many people does it impact? (2) Is there a conflict there? (3) Is it unusual? (4) Is it contrarian? (5) Is there some outrage factor, regardless of who would be outraged?

"After a story passes muster, you either end up with egg on your face and you lose your job, or you get sued if you screw up, but there are lots of well-written stories where you don't end up with egg on your face or sued."

Q: How and where do you draw the line between advocacy and environment reporting?

A: A member of the Sierra Club when in college and later a member of the Nature Conservancy, Rogers dropped his memberships deciding, as a newspaper reporter covering the environment, that, "my allegiance is to the truth."

"As long as you're hanging out the shingle that you're a newspaper reporter, you're not working for the Sierra Club. They have plenty of P.R. people.

"If you're a journalist and not an advocate, you should be asking the same questions of the environmentalists as the polluters, and keep the whole thing honest. If you're going to do fair journalism, you have to at least try to be objective, to show balance. Your credibility depends on asking difficult questions to all sides, and the value of the conversation, the very integrity of the democratic process depends on you as a reporter holding all sides accountable. Nobody has a monopoly on the truth.

"If I come in here with a Sierra Club t-shirt on, all of my stories are going to be buried. I want people to think that what I'm doing is serious journalism, and fair journalism, and that I am not a player on the stage. That like other journalists, I am a participant and a chronicler of what's happening on stage. Then the editors and readers trust me and will devote resources to do it more.

"In a democratic society, we have this wonderful marketplace of ideas where everybody gets to throw in their opinion. It's my job to give those opinions a bullhorn, to let the public hear those opinions evenly and fairly, and let the public decide. Information cannot be withheld when it disproves a point. You haven't given folks context unless you show folks basic numbers. As long as you include context, then you're providing a service to readers."

"There is a difference between investigative reporting and advocacy journalism. Investigative journalism is journalism with depth, context, and most importantly a foundation of factual, provable, empirical evidence. Advocacy is journalism trying to convince people of a viewpoint without those things. To me, advocacy is public relations. It's central function is to convince people of something, regardless of whether the facts back it up.

"Environment reporters have to ask themselves a few basic questions with every story: Why are you doing this? Is it because you see yourself as part of the environmental movement communicating to the public key issues that you think need reform? Or is your allegiance not to any movement but to the truth, even when it makes environmentalists look bad? Are you a journalist first? Advocacy journalists are the former, I am the latter.

"I have an opinion about every story I write. I ask who seems like they are carrying the weight of the evidence in this argument. So when I sit down to write, I ask myself: Who do I feel more sympathetic to? I might do

55 to 60 percent to the side I disagree with, just to make sure that in a story I'm not actually coloring it, or showing some bias. I also choose the most powerful, plain-spoken quote I can find."

Q: What are some realities of the environment beat?

A: "There are so many things on the beat to write, it's overwhelming," Rogers says. "I get more than 200 e-mails per day. I get stacks of faxes – at least 100 per day. I get at least 30 to 40 phone calls a day. My mail is delivered in those big plastic bins that the Post Office uses. I have 246 story ideas on my to-do list right now, so the world could stop turning and I could fill the paper with environment stories. There is this fire hose of stories that come and trying to figure out which one to take a drink off of..." Rogers sighs.

"A few reporters get a similar quantity of mail but most do not. It's because this beat is so broad. That mail is not people writing necessarily, it's the vast number of catalogs and magazines and reports and newsletters and things like that you have get on this beat. I get everything. I get everything from the Wilderness Society magazine to the Natural Resources Defense Council magazine. I get *Mining Voice.* I get cattle grazing magazines. I get the newspaper that goes to the Department of Interior employees, because if I'm going to try to understand what the main stories are from all viewpoints, I have to read all viewpoints. And so there's an immense amount of information that can be overwhelming. So I don't have a lot of time usually to do national stories unless there's some effect directly on the Bay Area.

"Environment reporting cannot be done well from a

desk. To understand the conversations, the reporter must go out and see the controversy, and talk to the people. It might change the theory you had in the beginning," Rogers said, describing his initial attitude when former President Clinton declared part of Southern Utah a national monument without even going there to do it. The declaration meant economic development in the form of a coal mine could not be located there.

Rogers recalls, "At first I was agreeing and sympathizing with the cowboys – saying it was a shameless political act, that he didn't even go to Utah to do it, he went to Arizona because he knew there would be huge protests where they were hoping to have a coal mine and other jobs in this poor area of Southern Utah where people have to drive 100 miles in each direction to go work at a motel. It's tough. And I thought, this is really low rent. This is tacky. He's playing to people in Hollywood, he can't even defend his own actions, and he's doing it by fiat because he knows Congress won't approve it. And I went out there, and I was standing at Escalante, Utah at sunset on the top of this mesa looking over this vast area that was the last place to get telephone service in the lower 48, and I said, well, they can't put a coal mine here. And I thought, tough s---- about those cattle ranchers because there are some values sometimes that trump local economic values."

Q: What does the future look like for environment reporting and for investigative reporting?

A: "I am optimistic about the future of environment reporting for several reasons. First, editors now accept that a separate person is needed to cover the environment. The beat is no longer considered frivolous or a novelty like it was in the beginning. Second, environmental journalism has matured.

"If you look at the early coverage, a lot of it was very activist. It was reprinting what the Sierra Club said without a lot of critical vetting. And in part because the problems were worse. When there was no Clean Water Act in 1969 and he Cuyahoga River caught on fire, it was pretty obvious what the problem was there. But when you're fighting about whether there should be 3 ppb or 1 ppb of something in a river full of fish and people swimming, it's a different story. The maturing of the beat has taken a generation, and has brought the same kind of healthy skepticism to this beat that is brought to all beats, and that's good for it.

"Concern for the environment is accepted as a mainstream value now. Industry also is recognizing environmental journalism as a serious and important pursuit. I am encouraged by what computer-assisted reporting has done for investigative reporting about the environment. People are doing a lot more sophisticated journalism than was done before. The amount of information that's out there online and the ability to crunch it is making environmental journalism much more sophisticated.

"To expand and nurture this kind of coverage, environmental journalists need to continue to increase their interaction with each other. That's first. It's a very lonely, difficult beat. It's a beat where it takes you a long time to learn the issues. You can spend your whole lifetime studying these issues and still not feel like you understand them. And it's a beat where a lot of people get mad at you no matter what you write. Therefore, it's a beat that has potentially high levels of burn-out, too. And it's a beat that – it's a kind of beat that lends itself nicely to having vetting sessions between reporters, where they can talk to each other about the trials and tribulations, about the things that work. They can see each other's work on e-mail listservs, and they can have quick conversations with each other via e-mail and other forums. It helps keep people encouraged. It helps teach people. It helps people feel good about what they do and feel that they're not alone out there. We have to keep emphasizing that it's hard, and you're going to get yelled at a lot, but it's worth it.

"But I think there still needs to be a very tough conversation about some of the underlying assumptions that environmental journalists have, and whether those are still valid. We need to have a conversation about what is the best way to clean the air? What is the best way to clean the water? How bad or good are things in the U.S.? Society has assumed, and reporters have assumed, that the government command and control model is always the best no matter what. But there are some examples where that's not true. You can say at a factory that we want you to reduce your output of smog by 50 percent and we're going to tell you how to do it. And most reporters say that's good environmentalism. And it's worked pretty well. The air quality has improved dramatically. But there's another model. There's, "We're going to tell you as a factory that we want smog reduced by 50 percent. We don't care how you get there. And by the way, anything additional that you do, we're going to allow you to sell those credits to someone else." So suddenly you unleash the creativity of those plants. You can do better than 50 percent because there's a reward for them.

"I'd also like to see more international coverage. We don't have pipes with red gunk running into streams any more. That's a good example where laws, environmentalism and activism have helped clean up America's waterways. We need, as journalists, to say, alright, let's not only make sure we write about the failures but also the successes. If most things are getting better in the U.S. and they are not getting better in the rest of the world, why aren't we reporting how bad they are in the rest of the world? It gets down to convincing editors that this stuff is relevant. You can write about loss of biodiversity, smog, clean water, fresh water, there are so many – fisheries, that's where the real downward trends and disasters are happening.

"I'd like to see us do all those things, but those things take money. They take open-mindedness. Those things take people challenging the status quo."

Rogers also notices another concern environment reporters will continue to confront in their work. Addressing the "How clean is clean?" question discussed earlier, he said, "Government is about the allocation of finite resources. If we spend an extra $500 million going from 2 parts per billion (ppb) to 1 ppb of arsenic, how many lives have we saved? How many cancers have we stopped? The general public doesn't want to ask that question, but it's because they don't necessarily understand risk factors."

Paul Rogers' Reporting Process

"At the *San Jose Mercury News*, the formal process for going about doing an investigative report involving the environment is:

(1) Propose the story at the beginning of the year

to the science editor, who is in charge of a 5-person team of reporters, including specialists in the environment, biotechnology, science, and a couple of medical reporters.

(2) The science editor pitches story ideas to the projects editor, who gives them a thumbs up or down.

(3) If accepted, the team budgets how much time they'll need, if the story is going to pull Rogers off his beat entirely, if its going to involve lawsuits for public records, database searches, and what kinds of resources the team will need, including travel.

(4) Once that is approved, the team sketches a thesis to try to prove. Rogers does that with his editor, and then begins exhaustive background research, making a list of main themes and ideas, potential sources, potential questions, trying out the thesis as he goes.

(5) During the process there is clear and regular oversight from editors. This is because its easy to sit in a cubicle and collect information forever. "It's always fun to collect it," Rogers says. "It's no fun to write it. It's pain. It's agony. You're opening your veins when you have to suddenly take stacks and stacks of notebooks that you spent months and months collecting information on. Hundreds of screens of database information. And start writing it and organizing it. That is agony. It doesn't matter how long you do it. It's hard. It takes a lot out of you. And so you need editors to hold your hand and ask you all along the way, what is this story about? What is this

story about? What do we have? What don't we know? What do we have to get now? Why does my grandmother care? Why is this relevant to the public? It helps keep the reporter focused."

(6) Write a first draft and turn it into the science editor, who carves it up.

(7) Rewrite until the science editor passes it on to the special projects editor for carving.

(8) Rewrite.

Then, after much yelling and screaming and pain, it gets in the paper, and sort of makes the world a better place," Rogers smiled and said.

Chapter 6: Dale Willman

When Ohio's Cuyahoga River caught fire in 1969, Dale Willman was growing up 90 minutes away. He wasn't reading newspapers, and he wasn't watching or listening to the news much either, although he watched Walter Cronkite because his parents did. The card-carrying Republican instead was busy trapping animals in the woods, and later writing a chess column for the Smithville (Ohio) High School newspaper, *Mosaic*. At 17 he became a disc jockey; but after a year, his boss suggested that he try a career in news instead.

Willman, an aspiring "nuclear propellerhead" majoring in international studies at Ohio State University, never completed his bachelor's degree, but did obtain a master's degree in environment and community from Antioch College New England. In 2002, after 30 years as a journalist, he dedicated the rest of his career to environment reporting by establishing Field Notes Productions, which provides stories about the environment for national and international radio broadcast.

Before becoming self-employed, Willman was a dishwasher, waiter, bagboy and night-mop-and-clean-up at a grocery store, summer landscaper, National Public Radio news anchor, reporter, editor and segment producer, Midwest editor of NPR's National Desk where he managed a staff of three plus a bureau chief and several freelancers, managing editor of the Great Lakes Radio Consortium, overseeing a $380,000 budget and staff of three serving a weekly news feed carried by more than 135 radio stations in 20 states, national correspondent/environment reporter for CNN Radio covering the White House, Pentagon, State Department and writing travel features; CBS radio correspondent; and Monitor Radio (Christian Science Monitor) anchor and senior editor. Along the way he landed prestigious journalism awards, twice winning the Edward R. Murrow Award – once for the best use of sound, and once for investigative reporting.

Willman also has taught journalism in Croatia and Macedonia, and served as the distinguished media studies professor at Carleton College. In addition, he trains professional journalists in storytelling, writing, and field technology for radio production.

Broadway's 'Beauty' leaves some performers feeling beastly: OSHA probes whether special effects harm health

CNN Radio, January 1, 1998
By Dale Willman

OSHA is investigating complaints from some actors and musicians that special effects used by Walt Disney

Co. in the Broadway production "Beauty and the Beast" are making them sick.

Inspectors with the federal Occupational Safety and Health Administration, armed with medical information from the musicians' union, made their first on-site visit last week to the Palace Theater, where the play opened in 1994. According to the union, fumes from the show's fireworks and fog are causing respiratory problems.

Since the play began, seven of the 24 musicians have been diagnosed with work-related respiratory problems. The latest complaint, a case of asthma, was confirmed in November. Last summer, one musician won a workman's compensation claim. Many actors say they frequently hold their breath on stage so they don't have to breathe fumes from the special effects.

Disney, while admitting the fumes may cause some "discomfort," denies that the chemicals in its theatrical pyrotechnics are harmful. The company says it has taken -- and will continue to take -- steps to make conditions more bearable.

OSHA initially relied on Disney's response to the charges. The company has work done on the theater's ventilation system, and commissioned a study to determine whether there was compliance with OSHA regulations. Disney said the study found no violations of federal law.

But environmental specialists who have investigated the workers' complaints say Disney's actions are inadequate, because the smoke could be creating permanent lung damage.

OSHA, which received the initial complaint from performers more than a year ago, has yet to make a ruling, and is unlikely to file a final report until later in 1998.

Workers say sickness showed up early

Almost from the day the show opened on Broadway in 1994, some of the performers said they began to get sick. By November 1995, several musicians were so concerned about their health that they asked for -- and received -- respirators from Disney's stage manager. Many continue to wear the masks at work today to filter the air before they breathe it.

One actor, who asked not to be named, said that on stage, the fumes dry out breathing passages.

"You try to manipulate your work so that you can hold your breath until you leave the stage, which is something a lot of us do," he said.

Environmental consultant Ed Olmstead was hired a year ago by the musicians' union to investigate workers' complaints. He found that the ventilation system drew air, including smoke and gas from special effects, from the stage directly into the orchestra pit.

A preliminary study conducted for the musicians' union by researchers at the prestigious Mount Sinai School of Medicine concluded that the show was to blame for the rash of respiratory illnesses. Medical histories of the performers showed that "they were doing fine until they began this assignment at 'Beauty and the Beast,'" said Dr. Jacqueline Moline, who conducted the study a year ago.

Disney: Airborne fumes an issue of discomfort

Disney Theatrical, which produces the "Beauty and the Beast" stage show, has said the performers may experience discomfort from the special effects, but are not being harmed. General Manager Alan Leevee said the show is so safe that he wouldn't mind sitting in the orchestra pit.

"We've maintained that the conditions in the Palace Theater conform with OSHA regulations, and that was our concern, that we met with the requirements dictated by OSHA," he said.

He said Disney has taken steps to improve ventilation systems in the theater, and has changed some of the show's special effects to try to make the musicians more comfortable.

However, respirators were not a part of the changes, Leevee said. While two stage managers said they had purchased respirators for musicians, Leevee said Disney never authorized such purchases.

He insisted that only the New York cast has complained about the effects. But the entire cast from a traveling show also complained about them in a letter to Disney more than a year ago, which they all signed.

Slow-moving OSHA investigation

Given the body of evidence supporting their claims, many "Beauty and the Beast" employees have started to ask why the government is not pursuing their claims more vigorously.

OSHA Assistant Area Director Brian Yellin said his agency was never given the medical data the musicians' union gathered. He also said that with OSHA's limited resources, the agency must give priority to fatalities and catastrophes.

Therefore, the organization's first on-site inspection took more than a year from the date the initial complaint was filed. OSHA's final report is not expected before next year.

Whether Disney actually meets federal regulations depends on how strictly OSHA, which watches worker health and safety, interprets its own guidelines.

Yellin said that while a company may meet the letter of the law, if a worker's health is in danger, a company like Disney could still fall under a rule known as the catch-all or general duty clause.

"The general duty clause requires an employer to provide a workplace free of recognized safety and health hazards," Yellin said. The word "recognized" in that clause is open to interpretation.

The pros and cons of firework secrecy

The performers have had no luck in trying to find out the composition of the fumes they are breathing.

M.P. Associates, the manufacturer of the special effects, refused to provide a list of the ingredients to OSHA for fear that their industry secrets will be stolen by other manufacturers.

Meanwhile, OSHA officials said they tried to subpoena the company to get the ingredients list, but the subpoenas were refused. It was unclear what subsequent efforts were made.

Thaine Morris, president of M.P. Associates, said the elements used to create the effects are safe, and that some could even be eaten without harm.

But industrial hygienist Monona Rossel of the Arts, Crafts and Theater Safety organization said that pyrotechnics manufacturers have never done any studies to find out what the explosions' byproducts are.

"You've turned these poor people, really, into lab rats," she said.

John Conklin of the American Pyrotechnics Association said fireworks can cause more than discomfort.

"There's no smoke that is designed for human inhalation, so any smoke can irritate the lungs," Conklin said.

Not even flashy effects may save Broadway

Disney and other Broadway producers may be reluctant to drop the glitz that special effects provide to stage shows like "Beauty and the Beast," because they are struggling to compete with the popularity of ever-more technical and flashy special effects in movies.

Yet a study this summer by a Boston-based consultant found that the popularity of Broadway is declining -- compared to television, movies, sports and gambling -- even with the pyrotechnics.

Instead of big bangs, Rossel advised, theater should return to the basics. And engineering consultant Harry Herman, who has studied the health effects of theatrical fogs, echoed her sentiments, saying Broadway should simply reemphasize good acting.

But Moline said the best reason to stop or at least change the effects is to preserve the health of those being exposed.

"Judging from what we know now," she said, "there should be a better way to have special effects in theaters that do not have any potential to have health effects for the workers, the performers, the stage hands, anyone in the theater environment who has to be there eight times a week doing their show."

Dale Willman on Environmental Journalism

Q: What drew you to environment reporting?

A: "To me, almost all environment reporting, because of the nature of the topics, has to be investigative. The reason is simple to me, and this is what attracts me to environment reporting: environment issues by nature are so complex. The environment covers so many disciplines. It's everything from architecture to macrobiology to ethnobotany to whatever. It's this crossing of disciplines so you're forced to be investigative in the sense that you really, in order to understand the issues that you cover, have to dig deeper. You have to do a lot of concentrating on the dots to make sense of those stories. That's part of the progression of my own growth.

"I started becoming interested in environment reporting at the same time I began to evolve in my understanding of journalism in general. I didn't place any value on [environment reporting, though]. I grew up in the [traditional journalism] school where you hear the voice of god and you impart the truth to people. I didn't realize how strongly I felt about the environment until I was much older. There were a lot of things that influenced me [such as family trips throughout the country to places like Dodge City and Yellowstone national Park]. It linked the environment to people. So I understood we were not separate from the environment, we were part of it."

Q: What personal lessons have you gleaned from investigative reporting about the environment?

A: "Am I an environmentalist? Yeah, I am. But do I try to advocate in my pieces? No." This clarity about his identity came from reporting about the environment, studying it, studying community, and completing an investigative report about the environment, he said.

Q: What makes the environment beat meaningful for you?

A: "For me, the clearest indication of a greater power is to be outside in the environment." [Describing living in Minnesota for a while, he said,] "The cold here is crisp, and the stars, on the nights when it drops down below zero and the clouds go away, I don't know how someone can say there's not a higher power. When I'm out hiking it's the same thing. For me, that's the greater experience." Reporting about the environment, he said, deepens his connection to the environment as well as its meaning.

Q: What makes investigative reporting meaningful for you?

A: "I have this incredibly strong, stubborn sense of fairness. I don't know where it came from, but I have this intense – I hate injustice. I hate abuse by the powerful. I get really mad about it. I used to get really angry in school about dumb things, about when people have power and they abuse it. I'm offended by that.

"To me, investigative reporting is looking at those who were abusing power and covering it up. It is a way of putting light on that and making people aware of an injustice. It's righting wrongs. It's this sense of anger and injustice that really pushed me to want to investigative reporting. The other element is that I also think investigative reporting is just really good reporting."

Q: How do you view investigative reporting?

A: "Before I ever did an investigative report, I had a clear idea of the impacts that kind of reporting had. At an early age, I learned Arizona Republic investigative reporter Don Bowles got blown up after uncovering organized crime. It was the story that brought about Investigative Reporters and Editors (IRE), and it got me thinking. Reporting about the Watergate break-in also intrigued me. But no opportunities to do it presented themselves. I became an early member of IRE when disco was popular, and started to learn about investigative reporting through the organization's conferences and newsletters, preparing for the opportunity.

"What turned me on to investigative reporting was in 'All the President's Men,' the eye of god above [Redford and Hoffman as Woodward and Bernstein] as they looked through the stuff at the Library of Congress. What turned me off to investigative reporting for a long time was the view that it was just laborious, and dreadfully long hours sitting in archives somewhere and just pouring over arcane documents. That was my view before I began to learn more about it, let alone do it.

"There's the view that all reporting is investigative in some sense. Which it is. And as I read more about it, my understanding grew. [I learned] it doesn't necessarily have to be records, but it is often prying things out that people don't want revealed – not in a '60 Minutes' 'Gotcha!' sort of way, but wading through things and finding that nugget, digging through and finding that one kernel that sends you into a direction and making connections.

"Document work to me is in a sense investigative reporting. It means connecting the dots. News reporting is often doing something about this dot, and that dot. And that dot over there. Investigative reporting and document reporting are connecting those dots. It's finding the threads. It's finding the commonality amongst those things."

Q: What criteria do you use for moving the public with your investigative reports?

A: "Being taken seriously is separate from hooking an audience. To me, the way to hooking people into a story is to be a storyteller. To present the story, at least in the beginning, through the eyes of someone affected by the circumstances. There's a basic convention in radio, and that is using a focus statement – one line that goes like this: Someone is doing something for a reason. In using that, that gives you a story that people can relate to. it puts it in human terms, and it gets people engaged, because chances are they probably know someone like that. So that's the way of hooking them in.

"How to move the public – that heads towards advocacy. I don't purposefully try to move the public in terms of a direction they should take. I try to move them to feel the story. To be a part of it. And to be compelled to do something. But I don't try to tell them what it is they should do."

Q: How and where do you draw the line between advocacy and environment reporting?

A: "There is a lot of advocacy in environmental journalism. I don't think it's wrong, but it's certainly not for me. There is so much baggage around the phrase environment reporting as it is, that as soon as people begin to feel you are an advocate, you lose a big chunk of your audience. While I respect a good many people who are doing advocacy journalism – I consider real advocacy journalism that which selects a point of view, and advocates for a certain point of view beyond all others. While there's a role for that, it's not what I want to do. You end up preaching to the converted rather than providing information for those undecided to make an informed choice.

"Is enviro reporting always advocacy? No. It does not need to be. You can take an issue and present that issue, and present the facts and allow the people to make up their minds. It, to some extent, rests on how you view the public's capacity to assimilate information. If you think they need to be persuaded to a point of view, you're going to do advocacy journalism. If you are convinced the public will make the right choice given enough information, then you'll just provide information rather than a point of view."

Q: Is investigative reporting about the environment advocacy?

A: "Is investigative reporting advocacy? That's a tough question. Can you be an environmentalist and be fair and objective? I think yeah, because you're human. Does the fact that you feel strongly about an issue mean that you're an advocate? To me, advocacy is propaganda whereas investigative reporting done well is not. The reporter may feel passionate about an issue, but works as hard as possible to be fair on that issue, and be as comprehensive as possible. One of the definitions of propaganda is to manipulate the truth to achieve a particular goal. And that means being selective in the information that you provide. To me, that's advocacy, and that's not straight journalism, the kind where you are informing people about issues significant to them.

"Just because you believe in something or understand something doesn't mean you're an advocate. Religion reporters proudly go to church or temple. Every business reporter I know is a capitalist. Does that mean they are biased in what they do? Yes! What I look at to determine if they are an advocate or not is the depth of their reporting. Your report doesn't necessarily aim toward a call for action, whereas advocacy does. I look at the weight of the evidence and make a judgment as to what is fair. For instance, with what we know today it's not fair to say that humans aren't at least partly responsible for global climate change.

"Many muckrakers in the past were advocates. I have a certain difficulty with that. Izzy Stone was a muckraker. People thought he was biased, but he dug out a lot of important stories. So I don't see much of a difference between investigative reporting and muckraking in that sense. Good muckraking to me is just good investigative reporting."

Willman views book-length journalism as the ultimate in comprehensive investigative reporting. Citing *Botany of Desire* by Michael Palen, about genetic engineering of crops, Willman said "He doesn't come out and say genetic engineering is bad. His purpose isn't to condemn genetic engineering. He made me consider all viewpoints and make up my own mind. The weight of the evidence does condemn genetic engineering.

"Economists say that in a market where there is complete information, people will make the right choices. I don't think that happens with environment issues. The Exxon Valdez oil spill was a huge plus for the gross national product (GNP) because we lost a little bit in fisheries but the hundreds of millions of dollars spent to clean it up actually helped the GNP. The loss of the vista, the loss of the fish that are a wonderful part of the ecosystem, the degradation of the ecosystem, that's not factored into the economic equation.

"I truly believe that if the public has complete information they are smart enough to make their minds up and it will be in favor of the environment. I don't think complete information is given to consumers about the environmental affects of their actions. And without that they cannot make informed choices. It's our jobs as journalists to give them that complete information, and then we should trust in their ability to make the right decisions. And that's what separates the journalism I believe in from advocacy journalism – advocates, it would seem, don't believe that people will make the right choice, so they must help them with that decision."

Q: What are some realities of the environment beat?

A: "What's different is the whole issue of the breadth of knowledge needed for environment reporting. That's the biggest reality. A separate reality is that many environment stories are slow to develop, which is counter to the news cycle that demands new, different, immediate. So journalists are trained to look for those sorts of stories – those which change rapidly, and with suddenness, rather than those that require a longer scale to put together to report on, and to get the public to understand. It makes it a harder sell to editors who are used to a whole different style of story.

"The environment is one of the most polarized beats. The language has been hijacked to a great extent so that it's hard to do a story without being accused of being an advocate or zealot for one side or the other."

Q: What does the future look like for environment reporting and for investigative reporting?

A: "I get angry at a lot of environment reporting now because I realize that [many of those doing it] just don't get it. They just don't understand it. Investigative reporting in part means providing greater context. It's that depth, that context that is so important for environment stories, and we don't have it. As a result, the public doesn't understand these issues, because they are ill-informed by the poor reporting being done by many people."

Dale Willman's Reporting Process

There is no rigid process for investigative reporting that Willman follows. The process for him is intuitive. One thing leads to another, he said. Throughout his life, once there was something he wanted to do, he set out to learn more about it. The same is true for stories, he said.

"Investigative reporting means a lot of conversations, calling people who know about a topic, calling people who understand it. I find I can't investigate something adequately until I understand it. I am kind of dense. So it means I have to gather a lot more information than probably I will ever use. But that context is important for giving me a greater understanding of the issue. That's the only way I can actually do an investigation."

He described the story about Disney's Broadway production of "Beauty and the Beast," for which he won the Edward R. Murrow Award for Investigative Reporting. The story revealed why the orchestra musicians wore respirators and double canister gas masks as they performed, which was because the ventilation for the production's pyrotechnics sucked harmful particulate matter into and through the orchestra pit. Willman picked up on the story three years into that practice and exposed it, working the story on his own time while employed by CNN Radio. To clarify, CNN Radio did not give him on-the-job time to investigative the story, but said they would be interested in running it if it passed scrutiny. CNN Television declined to do the story when Willman could not deliver tape, which happened because Disney got wind of the story before he could get the hidden camera video CNN Television wanted.

"The story took almost a year to collect. It took several months just to get permission from the musicians' union to go into the pit with the orchestra. I met with the union representatives, and the musicians, explained to them what I wanted to do, why I thought it was important, and that I was not interested in doing a story that would hurt them. I guaranteed any of them anonymity if they needed it, but said it was important that I had their participation. I told them I would not do the story unless I had full approval of everyone in the orchestra pit. [As he cultivated each union member involved,] I researched pyrotechnics, a business 'that is really cloooosed' because it mostly consists of family operations, which do not have public records. And regulators don't want to talk to you," he said.

Based on his research, Willman explained to me the chemistry that makes propylene glycol come out in a mist or fog on the stage, which floats the particulate matter into the air. The chemical is safe in foods, for ingestion, he said, but "no one knows what it does when it is inhaled, hot or cold," he said. Willman cited a case where a singer in the stage wings accidentally got a blast in the face of hot propylene glycol, which seared his lungs and permanently scarred his throat.

"I had very little support from CNN Radio for the story. The process in this story is a fairly unusual one because of having to get the approval and support of those who were in the pit. If they didn't support the effort, they could do a lot to screw it up. They could tell Disney we were doing the story. I just though it was the human thing to do, to make sure people agree this is the right thing to do. Many of the musicians didn't think it was a problem, didn't think it was effecting them. I had to convince them that some pyrotechnics have heavy metals in them because they add color, and that those metals have a huge effect on your health."

The research not only involved talking with the musicians, but also the fire marshals' association, regulators, legislative officials, a special interest organization aimed at protecting the health of performers in relations to their work, and plenty of reading and Internet research.

"This is exactly what I imagined investigative reporting to be," Willman said. "The people element of it was the part that was a little different. To have a large group of people I had to convince. But the rest of it really fit what I thought investigative reporting should be."

Appendix I: Timeline

Timeline of People and Events Shaping Environmental Consciousness in the United States 1945-1991.

Year	Milestone
1845	To avoid the waste and destruction of modern life, Henry David Thoreau withdraws to a cabin in the woods and later publishes Walden, a record of his thoughts and observations.
1849	U.S. Department of the Interior is established.
1857	Frederick Law Olmsted commissioned to develop the first great city park, New York's Central Park, introducing landscape architecture in the United States.
1864	George Perkins Marsh publishes *Man and nature.*
1869	John Wesley Powell leads first party to navigate the Colorado River through the Grand Canyon.
1872	First national park is established in Yellowstone, Wyoming.
1878	*Report on the lands of the arid regions of the west* by John Wesley Powell is released.
1879	U.S. Geological Survey is formed.
1882	The first hydroelectric plant opens on the Fox River in Wisconsin.

1890	The Census department declared the Frontier boundary, beyond which there were no more than two settlers per square mile.
1891	The Forest Reserve Act (March 3rd) permits federal government to set aside public land as forest preserve (precursor of the national forests). Yosemite National Park is established on October 1st.
1892	Sierra Club, an environmentalist club is founded by John Muir, pioneer in "aesthetic" conservation movement. The book, *Animal rights considered in relation to social progress* by Henry S. Salt is released.
1895	American Scenic and Historic Preservation Society is founded in response to industrialization and its damaging affects on the environment and on historical sites.
1897	Rise of Progressive Environmentalism, which leads to government intervention in the public interest in order to offset exploitation of natural resources by private developers.
1898	Cornell offers first college program in forestry. The Rivers and Harbors Act bans pollution of navigable waters.
1900	Lacey Act makes it a federal crime to transport illegally killed game across state lines.
1902	Bureau of Reclamation is formed and is responsible to protect water and other such resources. This propelled the Federal Land Reclamation program, which sought to return land that was once mined back to a natural condition.
1905	National Audubon Society is founded. Their original focus was to protect American birds as many were being killed for their feathers.
1907	Gifford Pinchot is appointed first chief of the U.S. Forest Service.
1908	The Grand Canyon is set aside as a national monument under the provisions of the Antiquities Act of 1906. Chlorination is first used extensively at United States water treatment plants, producing water ten times purer than when filtered. President Theodore Roosevelt hosts the first Governors' Conference on Conservation.

1913	The Hetch Hetchy Valley Dam, which would provide water for San Francisco, wins congressional approval after a five-year battle as building the dam would cause flooding in a scenic area in Yosemite.
1914	Martha, the last of the passenger pigeon species, dies in the Cincinnati zoo and later becomes a symbol for crusades against species extinction.
1916	The National Park Service, which seeks to preserve national parks, is established.
1918	Save-the-Redwoods League is created and the hunting of migratory birds is restricted by treaty between the United States and Canada.
1918	Save-the-Redwoods League is created and the hunting of migratory birds is restricted by treaty between the United States and Canada.
1918	Save-the-Redwoods League is created and the hunting of migratory birds is restricted by treaty between the United States and Canada.
1922	Izaak Walton League is organized. It is made up of hunters and others interested in conservation and protecting the environment in the United States.
1924	Environmentalist Aldo Leopold wins designation of Gila National Forest, New Mexico, as the first extensive wilderness area. The American conservation movement occurs with First National Conference on Outdoor Recreation.
1928	Boulder Canyon Project (Hoover Dam) is authorized to provide combined irrigation, electric power, and flood control system for the Arizona-Nevada border.
1930	Chlorofluorocarbons are hailed as safe refrigerants because of their non-toxic and non-combustible properties.
1933	The Tennessee Valley Authority is formed to analyze the environmental impact of hydropower projects before developing plans to harness the resources of the Tennessee River. The Civilian Conservation Corps employs more than two million Americans in forestry, flood control, soil erosion, and beautification projects in an attempt to boost the economy while addressing the needs of the land.

1934	The greatest drought in United States history is recorded. The Taylor Grazing Act regulates grazing on federal lands.
1935	The Soil Conservation Service is established. Extends federal involvement with erosion control. Wilderness Society is founded to expand and protect the nation's wilderness areas.
1936	National Wildlife Federation is formed. The national flood prevention policy is established by Omnibus Flood Control Act.
1940	U.S. Fish and Wildlife Service consolidate federal activities in wildlife management.
1946	Creation of the U.S. Bureau of Land Management centralizes administration of lands in the public domain. The Atomic Energy Commission is created to oversee the development of both peaceful and military uses of nuclear power.
1948	Air pollution incident in Donora, Pennsylvania, kills 20 people and causes 14,000 to become ill.
1959	The St. Lawrence Seaway, a joint Canadian and United States project, is completed, connecting the Atlantic Ocean to the western Great Lakes and providing 9,500 miles of navigable waters.
1961	Secretary of the Interior Stewart Udall articulates emerging ideas of humanity's ethical responsibility to preserve the environment as opposed to merely regulating the use of its resources.
1962	Silent Spring, an investigation of the dangers of unchecked pesticide use to the balance of nature, is published by Rachel Carson.
1963	First Clean Air Act authorizes $95 million to local, state, and national air pollution control efforts. Nuclear Test Ban Treaty between the United States and U.S.S.R. stops atmospheric testing of nuclear weapons.
1964	Wilderness Act creates National Wilderness Preservation System.
1965	Water Quality Act gives the federal government power to set water standards in absence of state action. The National Conference on Natural Beauty attacks the "uglification" of urban America and promotes the aesthetic rather than what would be more economic.

1966	Eighty people die in New York City from air pollution-related causes during a four-day atmospheric inversion.
1967	The Environmental Defense Fund is formed to lead the effort to save the osprey, a fish-eating hawk, from DDT.
1969	An oil spill in Santa Barbara, California, ruins beaches and focuses national interest on the growing pollution issues. In Alaska, $900,220,590 worth of lease bonuses on oil fields are sold in one day. Greenpeace is created, as Americans and Canadians join forces to protest nuclear bomb testing by the United States. *Design with Nature* by Ian McHarg advocates letting nature set design constraints on human decisions.
1970	The first Earth Day is celebrated on April 22. National Environmental Policy Act passes, requiring every Federal agency to issue an environmental impact statement for any dam, highway, or other large construction project under-taken, regulated, or funded by the Federal government. The Environmental Protection Agency is established to research, monitor, and enforce environmental laws and issues. The Clean Air Act Amends of 1963 toughens anti-pollution laws but fails to address acid rain and airborne toxic chemicals. The Natural Resources Defense Council is established as a combination of lawyers and scientists to develop policies of natural resource management.
1972	Use of DDT is phased out in the U.S. Federal Water Pollution Control Act (Clean Water Act) is passed, with the goal of restoring polluted waters for recreational use and eliminating discharges of pollutants into navigable waters. Representatives of 113 nations gather at the U.N. Conference on the Human Environment in Stockholm to develop a plan for international action to protect the world environment. Oregon passes the first bottle-recycling law. The U.S. Supreme Court decision supports the Sierra Club over Walt Disney Enterprises in a legal battle over use of Mineral King Valley, California. The Coastal Zone management Act, the Federal Environmental Pesticide Control Act, and the Ocean Dumping Act pass. The Club of Rome issues "The Limits of Growth", which provokes heated debate worldwide. This article examined many global issues such as the depletion of nonrenewable sources, the growth of industri-alization, and the deteriorating state of the environment.

1973	E.F. Schumacher publishes *Small is beautiful*, which is a book about ecological economics. The Convention on International Trade in Endangered Species of Wild Fauna and Flora (CITES) is signed by more than 80 nations. It is often called the "Magna Carta for Wildlife." In response to CITES, the United States passes broad-based Endangered Species Act, which applies to habitats as well as animal life. Controversy erupts over protection of an endangered fish called the snail darter during construction of the Tellico Dam in Tennessee, resulting in changes that weaken the Endangered Species Act. Congress approves licensing of 789-mile pipeline from Alaska North Slope oil field to Port of Valdez. The Arab oil embargo creates an energy crisis in United States.
1974	The Safe Drinking Water Act requires the EPA to set standards and policies to protect nation's drinking water.
1975	The last of the rivers in Tennessee is dammed. After a 100 year absence, Atlantic salmon return to the Connecticut River to spawn.
1976	The National Academy of Sciences reports that chlorofluorocarbon gases from spray cans are damaging the ozone layer. The Resource Conservation and Recovery Act (RCRA) empowers the EPA to regulate the disposal and treatment of municipal solid and hazardous wastes.
1977	The Department of Energy is created as a cabinet level office.
1978	Love Canal, New York, is evacuated after discovery that it sits on top of a chemical waste dump. Rainfall in Wheeling, West Virginia, is measured at a pH of 2, which is the highest level of acid yet recorded and 5,000 times more acidic than normal.
1979	Three Mile Island Nuclear Power Plant in Pennsylvania experiences near-meltdown. *Gaia: A new look at life on earth* is published by James E. Lovelock, who proposes that the Earth is a self-regulating entity, unconsciously maintaining the optimal conditions for life.

1980	The Comprehensive Environmental Response, Compensation, and Liability Act (Superfund) legislation is passed, requiring the EPA to supervise and regulate abandoned toxic waste site clean-ups. The debt-for-nature swap idea is proposed by Thomas E. Lovejoy, under which nations could convert debt into cash that would then be used to purchase parcels of tropical rainforest to be managed by local conservation groups. *Global 2000 report to the President* addresses world trends in population growth, natural resource use, and the environment by the end of the century, and calls for international cooperation in solving problems.
1981	Quebec Ministry of the Environment notifies the EPA that 60 percent of the sulfur dioxide pollution damaging its air and waters comes from industrial sources in the United States. The radical environmental action group EarthFirst! resorts to "ecotage" to gain objectives. "Ecotage" is sabotage done in order to protect the environment. The actions are done in secret so that there is no confrontation with police or other authority.
1982	The World Resources Institute is founded as an independent research and policy organization to help public and private groups pursue sustainable development.
1983	Interior Secretary James Watt resigns after overseeing an era of increased development of public lands and reduced concern for environmental consequences.
1986	Catastrophic failure of a Soviet nuclear power plant in Chernobyl contaminates large areas of northern Europe, mobilizes anti-nuclear forces, and stimulates the United States to undertake study of its federal nuclear facilities. Levels of dioxin 100 times the emergency level are found in town of Times Beach, Missouri, leading to evacuation and buy-out by EPA to allow decontamination.
1987	The *Mobro,* a Long Island garbage barge, travels 6,000 miles in search of a place to dump its trash, becoming a symbol of the nation's waste problems. The Montreal Protocol is signed by 24 countries pledging to halve chlorofluorocarbon production and use phased out by 1999, and later amended to require phasing out of CFCs by 1999.

1988	Plastic Pollution Research and Control Act bans ocean dumping of plastic materials. NASA scientist James Hansen warns Congress of global warming problem, saying that "greenhouse effect" may increase drought, melt polar ice, and raise sea levels. In response to discovery of widespread radon gas contamination of homes in the United States, the EPA study finds that indoor air can be 100 times as polluted as outdoor air. Beaches along the East Coast of the United States, Lake Michigan, and lake Erie are closed due to contamination by medical waste washed ashore. The Ocean Dumping Ban sets international legislation on the disposal of wastes into oceans.
1989	The New York Department of Environmental Conservation announces that 25 percent of the lakes and ponds in the Adirondacks are too acidic to support fish. Congress votes to halt timber sales in Alaska's Tongass National Forest, the last undisturbed rain forest in the United States. The Exxon oil tanker Valdez runs aground in Prince William Sound, Alaska, spilling 11 million gallons of oil in one of the world's most fragile ecosystems.
1990	Congress extends its ban on offshore oil drilling to cover 84 million acres of California, Alaska, and the East Coast. The United Nations report forecasts that world temperatures could rise 2 degrees Fahrenheit within 35 years because of greenhouse gas emissions. They warn that offending emissions must be reduced by 60 percent just to stabilize the atmosphere at the current level. The Clean Air Act amendments include requirements to control the emission of sulfur dioxide and nitrogen oxides into the air.
1991	War in Kuwait emphasizes the United States' dependence on imported oil and underscores the environmental damages of war. The United States accepts an agreement on Antarctica that prohibits activities relating to mineral resources, protects native species of flora and fauna, and limits tourism and marine pollution. Despite an espoused "no net loss" of wetlands, the Bush administration redefines the degree of standing water necessary for lands to be deemed as wetlands. This made it possible that 10 percent of the 100 million acres could be opened for development.

Source: World Resources Institute 1992 Environmental Almanac

Appendix II: Pulitzer Winners

Pulitzer Prize Winners for Reporting About the Environment

Year	Category	Description
1971	Public Service	*Winston-Salem (N.C.) Journal and Sentinel.* For coverage of environmental problems, as exemplified by a successful campaign to block a strip mining operation that would have caused irreparable damage to the hill country of Northwest North Carolina.
1979	National Reporting	James Risser of *Des Moines Register.* For a series on farming damage to the environment.
1992	Public Service	*Sacremento (Calif.) Bee.* For "The Sierra in Peril," reported by Tom Knudson who examined environmental threats and damage to the Sierra Nevada mountain range in California.
	Explanatory Journalism	James O'Byrne, Mark Schleifstein, and G. Andrew Boyd of *The Times-Picayune,* New Orleans, Louisiana for "Louisiana in Peril," which were articles about the toxic waste and pollution that threaten the future of the state.

1996	Public Service Editorial Writing	*News & Observer,* Raleigh, North Carolina for the work of Melanie Sill, Pat Stith, and Joby Warrick. They reported on the environmental and health risks of waste disposal systems used in North Carolina's growing hog industry. Robert B. Semple, Jr. of *The New York Times* for his editorials on environmental issues.
1998	Investigative Reporting	Gary Cohn and Will England of *The Baltimore Sun* for their compelling series on the international ship breaking industry. They revealed the dangers posed to workers and the environment when discarded ships are dismantled.

Source: The Pulitzer Prizes (online)

Appendix III: Pulitzer Finalists

Pulitzer Prize Finalists for Reporting About the Environment

Year	Category	Description
1980	National Reporting	The staff of the *Los Angeles Times* for its series on chemicals in the environment, "Poisoning of America."
1985	Editorial Writing	Jane Healy of *The Orlando Sentinel* for her editorials on Florida's environmental problems.
1994	Explanatory Journalism	The staff of *Newsday*, Long Island for its exhaustive investigation of breast cancer in the community, which included a probe of the environmental factors that may contribute to its spread.
	Feature Photography	Stan Grossfeld of *Boston Globe* for "The Exhausted Earth," a year-long series depicting the social, medical, and environmental crises caused by the depletion of natural resources.
1998	Beat Reporting	Keith Bradshaw of *The New York Times* for his reporting that disclosed safety and environmental problems posed by sport utility vehicles and other light trucks.

1999	National Reporting	The staff of *The Times-Picayune*, New Orleans, Louisiana for a revealing series on the destruction of housing and the threat to the environment posed by the Formosan termite.
2000	Breaking News Reporting	The staff of *The Oregonian* for its comprehensive coverage on an environmental disaster created when a cargo ship carrying heavy fuels ran aground and broke apart and how officials failed to contain the damage.
2004	Investigative Reporting	David Ottaway and Joe Stephens of *The Washington Post* for their detailed stories revealing dubious practices by The Nature Conservancy that produced sweeping reforms.

Source: The Pulitzer Prizes (online)

Appendix IV: Glossary

DDT – Dichlorodiphenyl trichloroethane (DDT) is a pesticide widely used in the United States on crops and to control insects that carry malaria and similar diseases. It is still used and recommended for use in Third World countries for malaria control.

Rachel Carson's 1962 book, *Silent Spring*, first reported how DDT accumulated in the food chain, causing cancer, genetic damage, and reproductive dysfunction among exposed species. Birds exposed to DDT laid eggs with weak, think shells that were easily broken, resulting in significant decreases in their population. In 1972, DDT was the first pesticide to be banned by the United States Environmental Protection Agency (West, et. al., 2003)

Dioxin – Dioxin has been called the most toxic chemical known to humans. In the 1970s it was linked to Agent Orange and adverse health effects among United States soldiers in combat in Vietnam. Most recently, the international community has called for the worldwide elimination of dioxin and other persistent organic pollutants. Dioxin is an umbrella name for a class of chemical compounds that contain carbon, hydrogen, oxygen, and chlorine. Dioxins are part of a larger class of compounds known as polycyclic halogenated aromatics. Dioxin is highly stable and insoluble in water. It adheres to clay and soot and dissolves in oil and organic solvents. Once released into the environment, dioxin remains for many years (West, et. al., 2003). A lethal dose of dioxin fits on the head of a pin.

<u>Chlorofluorocarbons</u> – Also known as CFCs, these are halocarbons, which are by-products of foam production, refrigeration, and air conditioning. In the stratosphere, CFCs contribute to the depletion of the earth's protective layer of "good ozone" (West, et. al., 2003).

<u>PCBs</u> – Stands for polychlorinated biphenyls, which persist in soil, sediment, water, waste disposal sites, and are found in some existing capacitors and transformers. Since they persist, they can be taken up from the soil by organisms and transferred through the food chain or accidentally leaked from electrical equipment. Their toxicity varies with the degree of chlorination and the actual position of the chlorine atoms on the basic structure. Some fish have been found to contain high levels of PCBs, which are transferred to humans when the fish are consumed. Poisoning from PCBs is associated with acne, respiratory distress, and liver damage (West, et al., 1995).

Genetic Modification – Genetic modification (GM) refers to the modifying of an organism, crop or animal by altering its DNA in some fashion, such as recombining its characteristics into a pattern not found in nature.

Global Climate Change – This concept refers to changes in weather patterns scientists have attributed primarily to air pollution and the decay of the earth's protective ozone layer.

References

Cates, J.A. (1997). *Journalism: A guide to Reference Literature* (2nd ed.). Englewood, CO: Libraries Unlimited.

Kovach, B. & Rosenstiel, T. (2001). *The Elements of Journalism.* New York: Crown.

Lippmann, W. (1931). "The press and public opinion." *Political Science Quarterly, 46,* 170.

Paneth, D. (1983). *The Encyclopedia of American Journalism.* New York: Facts on File.

The Pulitzer Prizes. (2005). http://www.pulitzer.org [electronic].

Stein, J. (1978). *The Random House Dictionary.* New York: Ballantine.

Webb, R.A. (1978). *The Washington Post Deskbook on Style* (Ed.). New York: McGraw-Hill.

Webster, N. (1981). *New Webster's Dictionary of the English Language.* (1981). New York: Delair.

West, B., Lewis, M., Greenberg, M., Sachsman, D., and Rogers, R. (2003). *The Reporter's Environmental Handbook.* (3rd Ed.). New Brunswick, NJ: Rutgers.

World Resources Institute (1992). *The 1992 Information Please Environmental Almanac.* Boston: Houghton Mifflin.

About the Author

Debra A. Schwartz, Ph.D., is a veteran freelance reporter based in the Chicago area and focusing on news and features involving science and the environment. She has written for numerous outlets including *Reuters, United Press International-Science, Wildlife Conservation, Environmental Science & Technology, Chemical Innovation, The Brownfields Report,* the *Chicago Tribune*, and *The New York Times.* She is also the author of *Listening Out of the Box: New Perspectives for the Workplace* (2004), and international award-winning research about intercultural listening in the workplace. This, her first book, concentrates on raw data from her research into whether investigative reporting about the environment is advocacy.

Apprentice House is the future of publishing...today. Using state-of-the-art technology and an experiential learning model of education, it publishes books in untraditional ways while teaching tomorrow's future editors and publishers.

Staffed by students, this non-profit activity of the Department of Communication at Loyola College in Maryland is part of an advanced elective course and overseen by the press's Director. When class is not in session, work on book projects is carried forward by a co-curricular organization, The Apprentice House Book Publishing Club, of which the press's Director also serves as Faculty Advisor.

Contributions are welcomed to sustain the press's work and are tax deductible to the fullest extent allowed by the IRS. For more information, see www.apprenticehouse.com.

Student Editors (2005-06)

Jerrell Cameron
Meghan Connolly
Katharine Dailey
Marie Guzowski
Natalie Joseph
Dana Kirkpatrick
Ann Marshall
Julia Sherrier
Marcus Smith
Joanna Walsh
Alison Wright
Kevin Zazzali

Printed in the United States
56433LVS00001B/61-108

9 781934 074015